氧化亚铜基可见光催化剂制备及性能研究

Preparation and Photocatalytic Properties of
Cu$_2$O-Based Visible-Light Photocatalysts

赵强 著

化学工业出版社

·北京·

内 容 简 介

本书主要介绍了纳米 Cu_2O 基复合材料光催化降解有机污染物的相关研究成果，对 Cu_2O 与其他半导体及不同材料复合后的光催化性能进行了探讨和研究，并对 Cu_2O 基纳米材料的光催化原理、制备方法及其在废水处理等相关应用进行了系统的研究。

本书可供从事光催化材料研究及废水处理的科研技术人员及高校功能材料、环境化学专业师生参考。

图书在版编目（CIP）数据

氧化亚铜基可见光催化剂制备及性能研究/赵强著.
—北京：化学工业出版社，2021.11（2023.8 重印）
ISBN 978-7-122-39915-1

Ⅰ.①氧… Ⅱ.①赵… Ⅲ.①氧化铜-光催化剂-制备-研究 Ⅳ.①O614.121

中国版本图书馆 CIP 数据核字（2021）第 188440 号

责任编辑：李晓红	文字编辑：王 琰
责任校对：边 涛	装帧设计：王晓宇

出版发行：化学工业出版社（北京市东城区青年湖南街 13 号　邮政编码 100011）
印　　装：北京建宏印刷有限公司
710mm×1000mm　1/16　印张 9　字数 152 千字　2023 年 8 月北京第 1 版第 3 次印刷

购书咨询：010-64518888　　　　　　　　　　　售后服务：010-64518899
网　　址：http://www.cip.com.cn

凡购买本书，如有缺损质量问题，本社销售中心负责调换。

定　　价：78.00 元　　　　　　　　　　　　　　版权所有　违者必究

PREFACE

环境污染和能源危机是当今人类面临的两大问题。其中，水污染是主要的环境问题之一。为了解决这一问题，光催化降解被认为是一种高效、环境友好的方法。尽管已经报道了许多关于 Cu_2O 光催化剂的研究，但基于 Cu_2O 的光催化剂的活性和稳定性仍然存在一些严重的问题：如氧化条件不稳定，光生空穴和电子的快速复合等。因此，本书的目的是通过与其他半导体形成异质结或者和其他具有大比表面积的导电材料负载来实现光生电子-空穴对在催化剂中的快速分离，从而开发高效且稳定的 Cu_2O 基光催化剂。此外，书中还将探讨有机污染物降解过程的光催化反应机理。

本书共分 6 章，其中第 1 章系统概述了 Cu_2O 及 Cu_2O 基复合材料制备方法的研究现状及其降解有机污染物的反应机理；第 2 章重点介绍了 Cu_2O 基光催化剂的制备及表征方法；第 3 章介绍了 p-n 型异质结 $g-C_3N_4/Cu_2O$ 光催化降解四环素的研究，通过对催化剂及反应条件的优化，探索提高光催化剂活性的有效途径和方法；第 4 章介绍了 Z 型异质结 $UiO-66-NH_2/Cu_2O$ 光催化降解甲基橙的研究，通过考察不同比例的 $UiO-66-NH_2$ 及反应条件的优化来探索提高光催化剂活性和稳定性的方法；第 5 章介绍了肖特基型异质结 Ti_3C_2Tx/Cu_2O 光催化降解四环素的研究，通过对催化剂及反应条件的优化，探索提高可见光光催化剂活性的途径和方法；第 6 章介绍了 Cu_2O 复合不同碳材料和多孔材料光催化剂的降解应用研究，通过对催化剂负载比例及反应条件的优化来研究具有大比表面积的导电材料在光催化剂中的应用。

第 3~5 章构建了不同类型的异质结结构，通过提高电子的捕捉位点，降低光生载流子的复合率；分别构建出 p-n 型、Z 型、肖特基型异质结结构，通过协同效应降低电子-空穴对的复合率，进而达到提高光催化降解四环素和甲基橙的性能和稳定性的目的。通过活性组分评价探讨反应过程中的活性组分，深入研究电子结构和界面结构对光生电子迁移机理和性能之间的关系，进而揭示光催化反应机理。

  本书得到了山西省面上基金项目（201901D111308）和山西大同大学博士科研启动基金的资助，在此表示感谢。

  由于笔者学识和经验有限，书中难免有疏漏之处，敬请读者批评指正，不胜感激。

赵　强

2021 年 8 月

CONTENTS

第1章 绪论 ·········001

1.1 引言 ·········002
1.2 Cu_2O 光催化剂 ·········003
1.2.1 具有规则多面体结构的纳米 Cu_2O 晶体 ·········003
1.2.2 中空纳米结构 Cu_2O ·········007
1.2.3 一维 Cu_2O 纳米晶体 ·········007
1.3 Cu_2O 基多相催化剂 ·········009
1.3.1 金属/Cu_2O 复合材料 ·········009
1.3.2 Cu_2O/金属氧化物复合材料 ·········015
1.3.3 Cu_2O/其他化合物 ·········023
1.4 结语 ·········033
参考文献 ·········033

第2章 催化剂的制备与表征方法 ·········050

2.1 试剂与仪器 ·········051
2.2 Cu_2O 的制备 ·········052
2.2.1 Cu_2O（菱方十二面体）的制备 ·········052
2.2.2 Cu_2O（立方体）的制备 ·········052
2.3 Cu_2O 基光催化剂的制备 ·········053
2.3.1 Cu_2O/EG 光催化剂的制备 ·········053
2.3.2 KAPs-B/Cu_2O 光催化剂的制备 ·········054
2.3.3 UiO-66-NH_2/Cu_2O 光催化剂的制备 ·········055
2.3.4 g-C_3N_4 纳米片/Cu_2O 光催化剂的制备 ·········055
2.3.5 Ti_3C_2Tx/Cu_2O 光催化剂的制备 ·········056
2.3.6 N-CMK-3/Cu_2O 光催化剂的制备 ·········057

2.3.7　Cu_2O/CNTs 光催化剂的制备 ·································· 058
 2.4　光催化剂的表征方法 ·· 058
  2.4.1　结构表征 ·· 058
  2.4.2　光催化性能评价 ·· 059
  2.4.3　活性组分 ·· 059

第 3 章　p-n 型异质结光催化剂 g-C_3N_4/Cu_2O ··············060

 3.1　引言 ·· 061
 3.2　g-C_3N_4/Cu_2O 催化剂的结构与性能表征 ···················· 062
 3.3　光催化降解四环素性能测试 ·································· 067
 3.4　光催化活性组分测试及机理研究 ······························ 068
 参考文献 ·· 070

第 4 章　Z 型异质结光催化剂 UiO-66-NH_2/Cu_2O ············074

 4.1　引言 ·· 075
 4.2　UiO-66-NH_2/Cu_2O 催化剂的表征分析 ······················ 076
 4.3　光催化降解甲基橙性能测试 ·································· 082
 4.4　光催化活性组分测试及机理研究 ······························ 083
 参考文献 ·· 085

第 5 章　肖特基型异质结光催化剂 Ti_3C_2Tx/Cu_2O ············088

 5.1　引言 ·· 089
 5.2　Ti_3C_2Tx/Cu_2O 催化剂的结构与性能表征 ···················· 091
 5.3　光催化降解四环素性能测试 ·································· 095
 5.4　光催化机理研究 ·· 096
 参考文献 ·· 098

第 6 章　Cu_2O 复合其他材料光催化剂 ·························102

 6.1　引言 ·· 103
 6.2　KAPs-B/Cu_2O 光催化剂光催化性能研究 ····················· 109
  6.2.1　KAPs-B/Cu_2O 光催化剂的表征分析 ······················ 109
  6.2.2　光催化性能测试 ·· 114
  6.2.3　光催化活性组分测试及机理研究 ····························116

  6.3 Cu_2O 复合碳材料的研究 ································· 117
    6.3.1 N-CMK-3 光催化剂光催化性能研究 ····················· 117
    6.3.2 Cu_2O/EG 光催化剂光催化性能研究 ······················ 124
    6.3.3 Cu_2O/CNTs 复合材料光催化性能研究 ··················· 127
参考文献 ··· 131

第1章

绪论

1.1 引言
1.2 Cu_2O 光催化剂
1.3 Cu_2O 基多相催化剂
1.4 结语

1.1 引言

环境污染和能源危机是当今社会人类面临的两大问题。水污染是主要的环境问题之一。半导体基光催化剂的光催化降解被认为是解决这一问题的一种高效、环境友好的方法[1,2]。TiO_2 用于光催化，取得了良好的效果。使用 TiO_2 的主要缺点是 TiO_2 作为光催化剂，它只有在紫外线照射下才有活性。因此，为了有效利用可见光，许多可见光半导体材料，如氧化亚铜（Cu_2O）、$BiVO_4$、Bi_2WO_6、氮化碳（$g-C_3N_4$）、Ag_2CO_3、WO_3、Ag_3PO_4 和 CdS 作为降解污染物的光催化剂已被研究[3-10]。其中，Cu_2O 以其低成本、环保、安全等优点吸引了大量的关注。它被广泛应用于传感[11-14]、光催化[15]、锂离子电池[16]、CO 氧化[17,18]和制氢[19]等多个领域。Cu_2O 是一种重要的 p 型半导体，直接带隙为 2.00~2.20eV，能吸收大部分可见光，其理论光电转换效率可达 18%[20,21]，是该领域一种有潜力的光催化材料。特别是，Cu_2O 作为一种降解有机污染物的光催化剂，受到了广泛的关注和深入的研究。图 1.1 给出了 Cu_2O 的结构类型及 Cu_2O 基光催化剂在光催化降解有机污染物的应用，下面几节将对它们进行详细介绍。

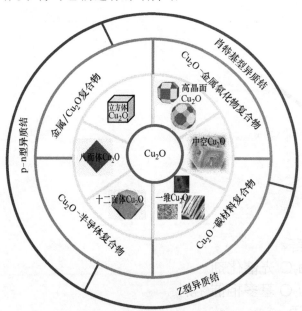

图 1.1 Cu_2O 基光催化剂种类及其作用机理

1.2 Cu_2O 光催化剂

在过去的几年里,人们通过各种方法制备出了不同形状和大小的纳米结构 Cu_2O,包括立方体[22,23]、八面体[24,25]、十二面体、多面体、中空结构[26]、纳米笼[27,28]和纳米线(NWs)[29,30]等形态,合成方法主要有共沉淀法、液相还原法、水热法、溶剂热法、电沉积法等。其中,共沉淀法在处理 Cu_2O 晶体的暴露面方面应用最为广泛[31-33],因为它在成核的不同取向和生长速率方面具有多种调控能力。众所周知,形状、尺寸和微观结构是决定纳米材料化学和物理性能的主要因素。

1.2.1 具有规则多面体结构的纳米 Cu_2O 晶体

(1) Cu_2O 纳米管

研究者对 Cu_2O 纳米管的合成进行了深入研究。Murphy 等人[33]采用液相合成法制备了密度均匀、单分散的 Cu_2O 立方纳米管。受表面活性剂浓度的影响,立方体的平均边缘长度在 200~450nm 范围变化。Kim 等人[34]采用多元醇法合成了 Cu_2O 纳米结构。他们发现,在聚乙烯基吡咯烷酮存在的情况下,只需要将硝酸铜与乙二醇在 140℃进行还原,就可以制备出多晶胶体球;之后引入少量氯化钠,得到了单晶纳米立方体。因此,他们认为氯离子在控制晶种的形成和各种晶体平面的生长速度中起着关键作用,进而使 Cu_2O 纳米结构变成纳米立方体。

Huang 等人[35]利用晶种介入的方法合成了尺寸为 40~420nm 的 Cu_2O 纳米立方体。他们发现十二烷基硫酸钠(SDS)表面活性剂等在制备过程中非常重要。这种晶种介入的方法可以很好地控制纳米立方体的大小和形状,然而这种方法的应用远远达不到工艺要求。此外,Chang 等人[36]开发了一种简单的一步成核控制方法来合成尺寸均匀、可调控的立方晶 Cu_2O,该方法产率可高达 87.5%。在此过程中,Cu^{2+}与柠檬酸离子螯合形成柠檬酸铜,可以延缓在加入 NaOH 过程中 $Cu(OH)_2$ 的沉淀速度。因此在初始阶段 Cu^{2+}可以显著降低 $Cu(OH)_2$ 生成,进而产生少量的 Cu_2O 晶种,从而增大 Cu_2O 纳米管的晶粒尺寸。所以，纳米立方体的尺寸随着柠檬酸钠浓度的增加而增大。

Nikam 等人[37]开发了一种在微波辐照下一步法合成 Cu_2O 的方法，溶液的 pH 值是关键的影响因素。在 pH=4.0 条件下，可以制备出不同尺寸的立方晶型 Cu_2O 纳米管。

Xu 等人[38]在室温下通过无添加剂的水热法制备了均匀的 Cu_2O 纳米颗粒，并在可见光下应用于甲基橙（MO）的降解。通过改变氢氧化钠溶液的浓度和铜盐的种类，可以在 20~500nm 范围内调节纳米立方体的大小。辐照 90min 后，尺寸较小的 Cu_2O 纳米管可使 MO 的浓度降低到 6%左右，较大的 Cu_2O 纳米管可使 MO 的浓度降低到 22%左右。这是因为晶体颗粒的大小不同导致晶体的比表面积不同，从而影响了催化剂的光催化活性。

Kumar 等人[39]通过一锅法合成了具有分层结构的 Cu_2O 纳米管，并在可见光下应用于降解亚甲基蓝（MB）染料的研究。在制备过程中，5nm 的 Cu_2O 纳米晶被葡萄糖自组装还原为 375nm 的分层结构。光催化实验表明，在可见光照射下，分层结构的 Cu_2O 纳米管比常规纳米管具有更强的光催化活性，这是由于 Cu_2O 纳米立方体提供了更多的表面活性位点并提高了其光电子性能。

Karthikeyan 等[40]通过简单的一锅法反应制备了尺寸可调且性能良好的 Cu_2O 纳米管。聚乙二醇（PEG）作为结构导向剂，用抗坏血酸还原 Cu（Ⅱ）盐可以得到 50~500nm 的结晶 Cu_2O 纳米管。在可见光照射下通过苯酚的降解来评价其光催化活性，结果表明将纳米管尺寸从 50nm 增加到 500nm，降解 6h 后苯酚的去除率从 36％提高到 50％。这项研究表明，较大的 Cu_2O 纳米管具有优异的苯酚矿化作用。

（2）八面体 Cu_2O 纳米晶体

许多研究人员也研究了八面体 Cu_2O 纳米晶体的合成。Wang 等人[41]在不使用任何表面活性剂的情况下，通过用肼还原氢氧化铜，制备了具有可调节边长的八面体 Cu_2O 晶体。他们比较了八面体 Cu_2O 晶体与立方体 Cu_2O 晶体对甲基橙（MO）的吸附和光降解的可见光光催化性能，结果表明暴露的 {111}晶面的 Cu_2O 八面体比 Cu_2O 立方体具有更高的活性。Choi 等人[42]通过电沉积法制备了八面体 Cu_2O 纳米晶体。首先通过电沉积获得带缺陷的立方晶体，然后通过含有$(NH_4)_2SO_4$ 的 $Cu(NO_3)_2$ 溶液第二次电沉积将其转化为八面体晶体。第二次电沉积的作用是填充立方晶体中的部分缺陷并调节{111}

晶面的生长。

Huang 等人[43]合成了具有从单分散截面立方体、立方八面体、截面八面体到八面体纳米晶体的系统形状演变的氧化亚铜纳米晶体。对于八面体纳米晶体，获得了尺寸为 160~460nm 的颗粒，通过光降解罗丹明 B 来评估其光催化的性能。实验结果表明，具有完全{111}表面的八面体 Cu_2O 纳米粒子比包含部分{100}和{110}表面的其他形状的纳米粒子在光催化反应中更有效。Lu 等人[44]还发现 Cu_2O 的催化活性受 Cu_2O 晶体形态的影响，对于罗丹明 B 的光降解，实验结果表明八面体 Cu_2O 的光催化活性最强，其次是截面立方体 Cu_2O 和八面体 Cu_2O。以制备的 Cu_2O 八面体作为光催化剂，在照射 180min 后 80% 的罗丹明 B 分子被降解。除了晶体颗粒的大小和晶体形态外，Cu_2O 的催化活性还与有机染料的电性能有关。

Huang 等人[45]制备了立方体、截面立方体、立方八面体等多种 Cu_2O 结构。研究表明，对于带负电荷的甲基橙的光降解，八面体和扩展六面体具有良好的催化活性，然而，只有{100}面的立方体没有活性。研究表明，电中性{100}面不能与带电分子很好地相互作用，因此不具有催化活性。对于光分解带正电荷的亚甲基蓝分子，立方体和八面体都无效，这是因为立方体和八面体不能很好地悬浮在亚甲基蓝溶液中；随着搅拌时间的增加，大量晶体逐渐漂移到溶液表面。这是因为带负电荷的分子可以更有效地吸附在 Cu_2O 晶体的{111}表面上，从而对甲基橙进行光降解。电中性{100}面不能与带电分子很好地相互作用并且催化失活。含有带正电分子的溶液可以排斥具有{111}表面的晶体，并使大量晶体漂浮到溶液的表面。因此，没有测量到{111}表面晶体的催化活性。

（3）十二面体 Cu_2O 纳米晶体

菱方十二面体代表了 Cu_2O 另一种重要的纳米晶体形态，因为 Cu_2O 晶体的晶面特性研究可以扩展到{110}面。Gao 等人[46]报道了菱方十二面体微晶的合成。在制备过程中，油酸作为表面选择性吸附添加剂，可以巧妙地决定最终晶体的形貌。使用不同浓度的油酸可以产生一系列的 Cu_2O 晶体形态，如立方体、八面体、{110}截面八面体和具有{110}表面的菱方十二面体。Yao 等人[47]制备了尺寸约为 100nm 的单晶十二面体 Cu_2O。

Huang 等人[48]在室温下合成了一系列具有从立方体到菱方十二面体结构的系统形状演变的 Cu_2O 纳米晶体。立方体、截面八面体、{100}菱方十二面体和菱方十二面体的平均尺寸分别约为 200nm、140nm、270nm 和 290nm。通过光降解甲基橙来评价其光催化性能，结果表明立方体基本上是无活性的。然而，在光照 90min 后，甲基橙能被菱方十二面体的 Cu_2O 完全光分解。研究者认为由于表面铜原子密度高，菱方十二面体仅暴露{110}面，因此对甲基橙的光降解表现出非常好的光催化活性。

（4）高晶面 Cu_2O

具有高晶面的纳米粒子也受到广泛关注，因为它在实际应用中可以表现出更高的化学活性。Zhang 等人[25]使用硬脂酸作为结构导向剂，通过水热法成功合成了完美的 26 面和 18 面 Cu_2O。在这个过程中，硬脂酸既是生长导向剂又是还原剂。他们研究了所制备的 26 面和 18 面多面体 Cu_2O 对甲基橙的吸附和光催化降解性能，并与八面体和立方体结构进行了比较，认为在 26 面和 18 面多面体的{110}面上具有更高的表面能和更高的"Cu"悬挂键密度使其具有更高的催化活性。

Leng 等人[18]基于控制溶液的方法制备了具有 50 面的多面体 Cu_2O 微晶，其产率高于 70%。在该制备过程中，混合溶剂中 OH^- 的浓度和极性有机溶剂与水的体积比对控制 Cu_2O 微晶的形貌起着至关重要的作用。Wang 等人[49]还通过调节 NaOH 的浓度，开发了一种简便的制备方法，在 $CuSO_4$/NaOH/抗坏血酸系统中对具有高指数面的多面体 Cu_2O 进行晶体修饰。首次制备了具有高指数{744}和{211}面的新型微米级 74 面多面体 Cu_2O。他们指出，一系列不同形态的 Cu_2O 晶体的形成应受动力学控制，最终合成得到的晶面取决于{100}和{111}面的不同生长速率。

（5）具有特殊结构的 Cu_2O 纳米晶体

Chen 等人[50]制备了一系列不同形式的 Cu_2O 纳米晶。结果表明，Cu_2O 纳米晶的特殊结构可以通过设计动力学控制路线代替传统的热力学控制来实现。Liu 等人[51]报道了在环境条件下通过 γ 辐射，以十六烷基三甲基溴化铵（CTAB）作为封端剂或模板，轻松合成亚微米到微米尺寸的单晶 Cu_2O 颗粒。通过控制 CTAB 与铜离子的浓度比，高产率地制备了具有特殊结构的 Cu_2O 纳米晶体，在此过程中，CTAB 作为成核和增长控制器是非常重要的。Shang 等人[52]提出了一种用于构建 3D Cu_2O 超结构的再结晶诱导自组装（RISA）工艺。结果表明，通过精确调整实验参数来平衡 CuCl 前驱体的水解和再结晶速率是成功的关键。

1.2.2　中空纳米结构 Cu_2O

中空纳米结构非常重要，因为它们具有高比表面积和足够的内部空间。通常，制备中空纳米结构有两种主要方法。最常见的方法是模板法，使用模板作为外壳后续生长的核心，之后再将该核心通过溶解、蚀刻或热处理来去除[53-55]。模板法制备过程很复杂。因此许多研究人员提出了自建无模板合成。Huang 等人[28]通过 $CuCl_2$ 为原料、十二烷基硫酸钠作为表面活性剂、盐酸羟胺作为还原剂、HCl 和 NaOH 调节水溶液的 pH，成功开发了一种合成截面菱方十二面体 Cu_2O 纳米笼等纳米结构的简便方法。Huang 等人[56]制备了带有蚀刻{110}面和中空内部的纳米结构，结果表明蚀刻能从表面{110}面快速进行到菱方十二面体的内部区域。Xu 等人[57]在室温下通过水溶液中的前体水解过程制备了单晶 Cu_2O 中空纳米立方体。通过调节反应物的浓度，可以获得边长为 50~200nm 的空心 Cu_2O 纳米立方体。结果表明，NaOH 和聚乙二醇（PEG）的添加是形成单晶空心结构的关键因素。

1.2.3　一维 Cu_2O 纳米晶体

一维（1D）纳米材料，例如纳米线（NWs）[58]、纳米管（NTs）[59]和纳米棒（NRs）[60]，由于其固有的各向异性以及电子和激子在最小维度内的有效传输，是非常有吸引力的器件构建模块。

（1）Cu_2O 纳米线

Tan 等人[61]开发了一种水热合成方法，用于生产具有可调直径和形态的单晶 Cu_2O 纳米线。首先，将 0.20g $Cu(Ac)_2$ 溶解在 40mL 去离子水中，向该溶液中加入 8mL 邻甲氧基苯胺水溶液，由于 Cu^{2+} 与邻甲氧基苯胺配位而得到深绿色溶液。然后，将反应混合物转移到 50mL 高压釜中并密封，温度在 140~180℃保持 5~10h，随后自然冷却至室温，最终得到 Cu_2O 纳米线。通过选择还原剂类型和合成温度，可以轻松调整 Cu_2O 纳米线的直径和形态。

Wang 等人[62]在室温下添加表面活性剂 PEG 后开发了一种新的还原方法，用于制备 Cu_2O 纳米线。首先将 PEG 和 $CuCl_2 \cdot 2H_2O$ 溶解在 H_2O 中，用磁力搅拌器搅拌至完全溶解，然后在搅拌条件下向上述溶液中滴加 NaOH，得到蓝色 $Cu(OH)_2$ 沉淀。搅拌 15min 后，加入水合肼溶液，由于水溶液中的 N_2H_4 是一种强还原剂，$Cu(OH)_2$ 沉淀逐渐被还原成红色的 Cu_2O 纳米线。TEM 图像表明，Cu_2O 纳米线的直径约为 8nm，长度范围为 10~20μm。

Ren 等人[63]在乙醇溶液中将有序介孔 Cu/Cu_2O 沉积得到多孔 Cu_2O 纳米线，得到的 Cu_2O 纳米线直径为 50~100nm，长度为 5~10mm。

（2）Cu_2O 纳米管

与 Cu_2O 纳米线相比，关于 Cu_2O 纳米管的报道相对较少。2003 年，Cao 等人[64]首次报道了 Cu_2O 纳米管的合成。在室温条件下，通过添加表面活性剂 CTAB（十六烷基三甲基溴化铵），用弱还原剂葡萄糖还原 $[Cu(OH)_4]^{2-}$ 得到 Cu_2O 纳米管。Jin 等人[65]通过 Fehling 反应，用葡萄糖或果糖将 Cu^{2+} 还原为 Cu^+，在水溶液中制备了 Cu_2O 纳米管。结果表明，在晶体生长开始时 Cu_2O 纳米管首先形成，但是随着时间的延长，最初空心的 NTs 可以被填充成实心的 NWs。与之前的研究不同的是，在这个过程中没有添加任何表面活性剂，因此，一维纳米材料的另一种生长机制得到了证实。

Zhong 等人[14]使用两步法制备了单晶六边形 Cu_2O 纳米管阵列，在 NH_4Cl 作用下形成管状结构。NH_4Cl 和 Cu_2O 形成可溶性的 $[Cu(NH_3)_4]^{2+}$。此外，溶液中的 Cl^- 也会导致沉积溶液中形成 $CuCl_4^{2-}$ 并吸附在 Cu_2O 纳米棒的表面。由于静电作用，这些吸附的 Cl^- 和 $CuCl_4^{2-}$ 为 Cu_2O 中的 Cu^+ 提供电子并增强 Cu—O 键。Cl^- 和 $CuCl_4^{2-}$ 离子吸附化学作用的特性将导致 Cu_2O 在纳米棒表面的溶解速度比纳米棒内部的溶解速度慢得多。因此，选择性溶解优先沿 c 轴影响 Cu_2O 纳米棒的中心部分，留下侧面并导致形成管状结构。

（3）Cu_2O 纳米棒

Cu_2O 纳米棒的制备是近年来才报道的。Guan 等人[66]在水-甲苯体系中，以水杨醛为配体和还原剂，通过界面蚀刻法制备了一种新型氧化亚铜纳米棒结构。水杨醛和 Cu^{2+} 可在水/油界面形成络合物，在水热条件下，络合物被蚀刻还原为具有特殊纳米棒结构的氧化亚铜。Musselman 等人[67]首次研发了一种通用技术，通过电沉积在导电基板上制备大面积、自组装、尺寸可控的纳米棒阵列。Lee 等人[68]在碱性条件下通过电化学沉积在阳极氧化铝（AAO）模板上制备了 Cu_2O 纳米棒。在室温下 Cu_2O 纳米棒的生长速率达到 360nm/min。结果表明，在恒流条件下先形成 Cu_2O 纳米管，然后填充得到

Cu_2O 纳米棒。Haynes 等人[69]通过新型模板化电沉积工艺制备了 Cu_2O 纳米棒阵列。首先在透明导电氧化物基板上制备作为牺牲模板的 ZnO 纳米棒薄膜，然后在 ZnO 纳米棒表面涂抹硝化纤维，聚乳酸作为纳米棒阵列之间的填充材料，然后通过蚀刻工艺去除 ZnO 获得 Cu_2O 纳米棒阵列。

总之，作为一种有应用前景的有机污染物降解光催化剂，Cu_2O 的物理和化学性质在不同的形态、尺寸和结构上存在明显差异。此外，Cu_2O 的催化活性还与有机染料的电学性质有关。Cu_2O 光催化降解有机污染物的研究主要集中在具有完整晶面的纳米晶体的应用上，一维 Cu_2O 降解有机污染物的研究较少，其光催化性能有待进一步研究。虽然随着研究理论和方法的发展，Cu_2O 的形貌、结构、粒径等都可以在微观层面进行调控，并取得了很大的进展。但是 Cu_2O 的窄带隙导致光生电子和空穴的快速复合，使 Cu_2O 的光催化活性受到限制[25]。而且光照射 Cu_2O 在水溶液中的稳定性差[19]，由于 Cu_2O 的价带（VB）低于 Cu 的氧化电位[70,71]，很容易被氧化为 CuO，从而限制了其光催化效率。

1.3　Cu_2O 基多相催化剂

为了进一步提高 Cu_2O 的光催化性能，将其与其他半导体复合被认为是克服这一缺点的有效方法。据报道，Cu_2O 可以与一些宽带隙半导体复合形成纳米复合材料，其具有良好的可见光响应和丰富的反应位点，从而提高 Cu_2O 的光催化效率[72,73]。另外，与窄带隙材料的杂化可以有效抑制光诱导电荷载流子的复合，这有助于提高 Cu_2O 的优异光催化性能[74,75]。迄今为止，随着纳米材料科学和纳米技术的飞速发展，研究者已经合理设计和合成了具有优异控制成分、形状和尺寸的 Cu_2O 基异质纳米结构，这可能为增加 Cu_2O 的潜在应用带来优良的性能。

1.3.1　金属/Cu_2O 复合材料

众所周知，由于在金属-半导体界面处形成肖特基势垒，可以有效地防止光催化过程中光电子-空穴对的复合，通过与金属结合可以显著提高 Cu_2O 半导体的光催化效率[76]。因此，金属/Cu_2O 纳米结构已经被广泛研究。

（1）Cu/Cu_2O 复合材料

铜是一种重要的金属，具有良好的高导电性和导热性[77,78]。铜基材料有着潜在的应用前景，人们已经尝试采用各种方法制备 Cu/Cu_2O 复合材料，如溶胶-凝胶技术、乳液法、水热合成法、热解法、化学气相沉积法等[79-81]。

Sun 等人[82]使用简单的液相还原法在 26 面体 Cu_2O 结构的{111}面上合成低成本 Cu 纳米颗粒，并用于甲基橙的光降解。如图 1.2 所示，该方法制备的 Cu_2O 为 26 面体，Cu 纳米颗粒成功地分布在 26 面体 Cu_2O 结构表面。与纯 Cu_2O 相比，新型 Cu/Cu_2O 异质结构对甲基橙具有更好的吸附和光降解。由密度泛函理论（DFT）计算可知，Cu_2O 的{111}表面比{110}和{110}表面更容易还原，最终导致 Cu 纳米粒子在 Cu_2O{111}表面上选择性生长。优异的光催化性能归因于 Cu 纳米颗粒的引入，Cu 纳米颗粒表现出更高的吸附能力和光催化活性，可增强紫外光下 MO 染料的降解。

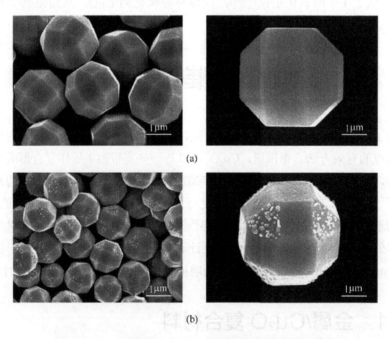

图 1.2　Cu_2O 和 Cu/Cu_2O 异质结构的 SEM 图[82]

(a) Cu_2O；(b) Cu/Cu_2O

Chen 等人[83]使用新型氧化/还原工艺制备了核壳异质结构的 Cu/Cu_2O 纳米线。可见光照射下核壳结构的 Cu/Cu_2O 纳米线在亚甲基蓝光降解中表现出优于其他对应纳米结构的光催化活性。光照120min后，Cu/Cu$_2$O NWs的溶液中仅保留了7.9%的初始染料，而Cu/Cu$_2$O纳米颗粒（NPs）和Cu$_2$O纳米簇（NCs）在光照同一时间后分别保留了23%和40%的初始染料。对于CuO NCs，亚甲基蓝在光照120min后有90.6%未降解。特殊的核壳异质结构和三维空间构型使电荷和空穴有效分离这正是Cu/Cu$_2$O NWs在可见光下表现出优异可见光活性的主要原因。

Ai等人[84]采用界面水热法制备了核壳型Cu@Cu$_2$O微球，在制备过程中首先通过Cu（Ⅱ）的还原形成纯铜微球，然后表面Cu被氧化转化为Cu$_2$O壳，形成Cu@Cu$_2$O核壳结构。对于气态一氧化氮在可见光照射下的降解，所制备的Cu@Cu$_2$O核壳微球与Cu$_2$O相比表现出增强的光催化活性。他们指出，这种增强的光催化活性可能归因于Cu@Cu$_2$O的核壳结构，它有利于电荷转移并抑制光生空穴-电子对的复合。

（2）Ag/Cu$_2$O复合材料

为了有效地利用阳光，在半导体表面沉积了贵金属等纳米结构，如Au、Ag和Pt。在复合材料中，金属充当电子接受体，引入有效的界面电荷分离，从而在扩展的波长范围内显著提高光催化活性[85-88]。Zhang等人[89]基于一锅法制备了Cu$_2$O/Ag复合纳米球（CNSs）。在制备过程中，将不同比例的AgNO$_3$溶液加入Cu$_2$O纳米球母液中，酸性溶液中的Cu$_2$O纳米球可还原Ag$^+$离子使Ag纳米颗粒直接沉积在Cu$_2$O纳米球表面。结果表明，可以通过控制AgNO$_3$溶液的体积来调节Cu$_2$O纳米球上Ag的含量。在可见光照射下通过光降解甲基橙研究了Cu$_2$O/Ag CNSs和纯Cu$_2$O的光催化活性。结果表明，由于增强了光吸收和电子效应，所获得的Cu$_2$O/Ag CNSs具有比Cu$_2$O和Ag纳米颗粒更优异的光催化活性。

如图1.3所示，Lee等人[90]使用简单的共还原法合成Ag/Cu$_2$O核壳纳米粒子，并研究了纳米粒子的壳厚度对其光催化活性和稳定性的影响。通过甲基橙降解测试光催化性能，发现随着壳厚度的增加，活性和稳定性同时提高。与纯Cu$_2$O相比，Ag/Cu$_2$O核壳纳米粒子显示出更宽的紫外-可见光谱吸收范围。活性的增强可归因于凹凸结构的高比表面积和来自Ag核的等离子体电荷转移。

Xiong等人[91]在室温下制备了一维等离子体Ag@Cu$_2$O核壳异质纳米线，并发现所得的一维Ag@Cu$_2$O NWs对有机污染物的降解表现出比Ag@Cu$_2$O优异的

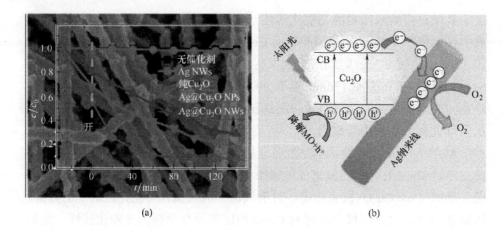

图 1.3　Ag/Cu$_2$O 异质结光催化降解实验（a）及机理（b）[90]

光催化活性。光催化活性的增强可归因于表面等离子体共振和银纳米线核的电子下沉效应，以及独特的一维核壳纳米结构。

Hu 等人[92]使用简便的湿化学方法制备了 Cu$_2$O 涂层的 Au/Ag 纳米棒（NRs），在可见光照射下通过甲基橙的降解来评价制备的具有不同 Cu$_2$O 厚度的 Au/Ag/Cu$_2$O NRs 的光催化活性。与球形 Ag/Cu$_2$O 纳米粒子相比，Au/Ag/Cu$_2$ONRs 表现出增强的光催化活性。此外，Au/Ag/Cu$_2$ONRs 的光催化活性取决于壳厚度，Cu$_2$O 壳的最佳厚度约为 20nm。增强机制归因于通过等离子体诱导的共振能量转移（PIRET）促进 Cu$_2$O 中的电子-空穴分离。

Sharma 等人[93]在室温下制备了三元 Ag/Cu$_2$O/rGONCs（rGO 为还原氧化石墨烯），该方法价格低廉，易于制备。在 Ag/Cu$_2$O/rGO NCs 中，rGO 的功能化还原显著改变了费米能级，这可能导致从 Cu$_2$O 到 rGO 的电子注入速率增强。

Wei 等人[94]通过一锅法制备了三元还原氧化石墨烯包裹八面体 Ag/Cu$_2$O/rGO 光催化剂。在复合材料中，用致密的 Ag 纳米粒子修饰 Cu$_2$O 八面体，rGO 纳米片包裹在其表面。可见光下对水中的苯酚光降解，三元 Ag/Cu$_2$O/rGO 光催化剂表现出比纯 Cu$_2$O、二元 Cu$_2$O/rGO 或 Ag/Cu$_2$O 高得多的活性。这是因为 rGO 纳米片的添加阻止了 Ag 纳米粒子团聚并提高了八面体 Ag/Cu$_2$O 结构的稳定性。

（3）Au/Cu$_2$O 复合材料

作为一种典型的金属和氧化物半导体材料，Au@Cu$_2$O 材料以其新颖的结构和多方面的潜在应用引起了广泛关注。如图 1.4 所示，Liu 等人[95]证明金纳米颗粒在截面八面体和立方八面体 Cu_2O 晶体的{111}表面上的选择性生长也可以实现。图 1.4 为不同标尺下 Au/Cu_2O 异质结构的 SEM 图，金纳米颗粒均匀地分布在八面体 Cu_2O 表面上。Au/Cu_2O 中 Au 纳米颗粒的密度和尺寸可以通过调节 Au 前驱体的浓度来控制。实验发现在 $HAuCl_4$ 的浓度较低时，大量 Au 纳米颗粒形成在 Cu_2O 晶体的{111}表面，但几乎没有 Au 形成在{100}表面；当增加 $HAuCl_4$ 的浓度时可以得到 Au/Cu_2O 核/壳异质结构。该 Au/Cu_2O 纳米复合材料对 H_2O_2 还原具有比纯 Cu_2O 晶体高 10 倍的电化学催化活性，催化作用的增强主要归因于界面处 Au NPs 的极化，这使得 Cu_2O 对 H_2O_2 还原活性更高。

图 1.4 不同标尺下 Au/Cu_2O 异质结构的 SEM 图[95]

(a) 800nm；(b) 400nm；(c) 300nm；(d) 500nm

Zhu 等人[96]开发了一种 Au 纳米颗粒在 Cu_2O 八面体表面特定位置选择性生长的策略，制备过程具有一系列形态演变。在室温下通过对甲基橙水溶液光催化降解来评价催化剂。光催化降解实验表明，与纯半导体纳米晶体相比，Au/Cu_2O 异质结构增强了光催化活性，提高了光诱导电荷分离效率，从而增强了光催化活性。

Mahmoud 等人[97]制备了具有不同纳米层厚度的 Cu_2O/Au 纳米结构并用于染料降解。结果表明等离子体 $Au@Cu_2O$ 核/壳异质结构可以带来许多优点,例如有效地保护金属纳米粒子免受腐蚀,通过金属核和半导体壳之间的三维空间使金属-载体相互作用最大化,控制中心波长局域表面等离子体共振。

Wang 等人[98]制备了 Au/Cu_2O 核壳异质结构,并研究了多面体金纳米晶体形态对其形成的影响。通过改变还原剂的体积获得核壳立方体、面凸起立方体、核壳八面体和凸八面体,并比较了它们对甲基橙降解的光催化活性。发现所有立方体都没有光催化活性,因为它们基本上被 {100} 面包围。具有更多{111}面的凸八面体显示出比常规 Au/Cu_2O 核壳八面体更好的光催化性能。

(4) Pd/Cu_2O 复合材料

研究人员还研究了 Pd/Cu_2O 纳米结构。Miller 等人[99]通过水热法制备了 Pd/Cu_2O 纳米结构(见图1.5),并用于在可见光照射下降解四溴二苯醚。获得的 Pd/Cu_2O 证明了可有效脱除 2,2′,4,4′-四溴二苯醚(BDE-47)分子中的溴,BDE-47 是环境问题最严重的多溴二苯醚(PBDE)之一,在可见光照射下,在反应 10h 内大约 70%的 BDE-47 被 Pd/Cu_2O 光催化剂脱溴降解为低溴化的同系物。这是首次报道的非二氧化钛基金属氧化物半导体用于光催化降解多溴二苯醚的案例。通过在 Cu_2O 表面掺入 Pd 纳米粒子,Pd/Cu_2O 能够激活生成的氢以完成 PBDE 的还原加氢脱卤。

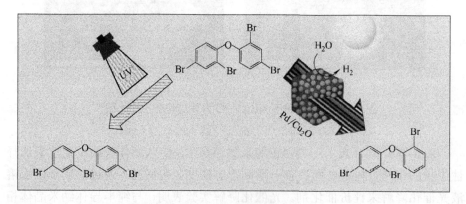

图1.5 Pd/Cu_2O 异质结构[99]

t（5）Zn/Cu$_2$O 复合材料

Zn 掺杂不仅影响 Cu$_2$O 晶体的形貌，而且对其光电性能也有显著影响。Heng 等人[100]以 CuWO$_4$ 为前驱体，通过原位光化学方法制备了具有中空微立方体形态的 Zn 掺杂 Cu$_2$O 光催化剂。实验结果表明，当 Cu$_2$O 中掺杂 0.1%（质量分数）的 Zn 时，光催化活性显著增强，表观量子产率为 38.95%。

Yu 等人[101]采用溶剂热法制备了 Zn 掺杂的 Cu$_2$O 颗粒，用于光催化降解环丙沙星（CIP），发现 Zn 掺杂可以有效提高 Cu$_2$O 的光催化性能（见图 1.6）。制备的 Zn 掺杂 Cu$_2$O 颗粒具有 {111} 晶面的择优取向，可以缩短光生载流子到达表面的时间，从而提高光生电子和空穴的传输和分离效率。此外，Zn 掺杂可以增加 Cu$_2$O 的带隙，提高电荷的传输和分离效率。

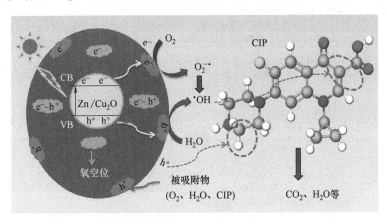

图 1.6　Zn/Cu$_2$O 异质结构光催化降解环丙沙星的机制[101]

1.3.2　Cu$_2$O/金属氧化物复合材料

Cu$_2$O 的电子容易被可见光激发，从而参与废水中染料的降解过程。然而，光生电子和空穴易复合会影响其光催化活性。Cu$_2$O 与金属氧化物的结合已成为克服上述光生电子与空穴复合缺点的有效手段。

（1）Cu$_2$O/ZnO 复合材料

氧化锌（ZnO）由于其成本低、光催化活性高和无毒性质而被公认为优选的光催化剂。然而，ZnO 由于其 3.2eV 的宽带隙，只能在紫外区工作。Cu$_2$O

作为 p 型半导体，窄带隙为 2.0eV，ZnO 作为 n 型半导体之一，能带的相对位置特别宽，其中 Cu_2O 的导带（CB）比 ZnO 稍高而 ZnO 的价带（VB）低于 Cu_2O。因此，这可能保证 Cu_2O/ZnO p-n 型异质结在光催化领域有价值的应用。

Teo 等人[80]通过共沉淀法制备了 Cu_2O/ZnO 纳米复合材料并应用于降解染料废水。结果表明，在最佳条件下，Cu_2O/ZnO 纳米复合材料在反应时间 2h 内实现了接近 80%的苏丹红脱色，该效率远高于纯 Cu_2O 或 ZnO。图 1.7 为 Cu_2O/ZnO 异质结构降解染料废水时的机理，可以看出当 p 型 Cu_2O 负载在 n 型 ZnO 半导体表面时形成 p-n 型异质结时，ZnO 的导带比 Cu_2O 的导带低 0.5eV[102]，因此 Cu_2O 中产生的电子可以转移到 ZnO 表面，使 Cu_2O 中的电子-空穴快速分离并避免复合，从而提高光催化活性。

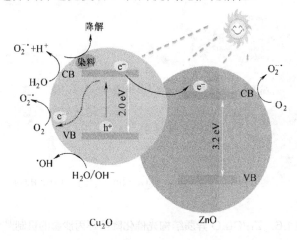

图 1.7　Cu_2O/ZnO 异质结构用于降解染料废水的机理[80]

Deo 等人[103]通过在 Cu_2O 纳米针上分层生长 ZnO 纳米棒，制备了 Cu_2O/ZnO 异质纳米结构。通过降解甲基橙染料来测试 Cu_2O 纳米针、Cu_2O/ZnO 异质纳米结构的光催化活性。与 Cu_2O 纳米针相比，由于在 p-Cu_2O/n-ZnO 界面处形成多个 p-n 型异质结以及两种材料的准一维结构，光催化性能显著增强，从而有效地进行了电荷分离和传输。纯 Cu_2O 纳米针降解甲基橙 2h 后，剩余的甲基橙量约为 70%，而在 Cu_2O/ZnO 异质结降解 2h 后，剩余的甲基橙量仅为 7%左右。

Zou 等人[104]通过两步法（水热法和电沉积法）在掺铟氧化锡（ITO）玻璃基板上制备了垂直排列的 Cu_2O/ZnO 异质纳米棒阵列。通过降解甲基橙来评价甲基橙的吸附和光催化降解。结果表明，在可见光照射下，纯 ZnO 纳米棒阵列对甲基橙的脱色能力非常弱，而 Cu_2O 薄膜对甲基橙的脱色能力很强，但光催化活性仍低于 Cu_2O/ZnO 异质纳米棒。Cu_2O 电沉积时间为 20min 制备的纳米棒阵列具有最高的光催化活性，可在 5h 内将甲基橙降低至 10% 左右。增强的光催化活性归因于 Cu_2O/ZnO 异质纳米棒阵列结构更好的光散射以及 Cu_2O/ZnO 异质结内电荷的快速分离和传输。

Kandjani 等人[105]还制备了 Cu_2O/ZnO 核/壳光催化剂，其中单分散的 Cu_2O 纳米立方体作为核，具有不同形态的 ZnO 纳米颗粒包覆为壳。结果表明，所制备的核/壳纳米颗粒的光催化效率不仅取决于 ZnO 和 Cu_2O 纳米颗粒之间 p-n 型异质结的形成，还取决于对壳形成的覆盖率和形态的控制。

Wu 等人[106]在 Cu_2O 立方体、八面体和菱方十二面体的表面制备了稀疏的 ZnO 纳米结构，评价它们的光催化活性，以证明对界面电荷转移的界面效应，并发现使用纯 Cu_2O 晶体沉积的 Cu_2O 菱方十二面体表现出预期的强催化活性，而 ZnO 沉积的 Cu_2O 立方体对光降解反应保持惰性，说明对光催化的负面影响。此外，具有光催化活性的 Cu_2O 八面体在 ZnO 纳米结构沉积后变得完全不活跃，是因为 ZnO 的 {101} 面优先生长在 Cu_2O 的 {111} 面上，导致催化活性的下降，因此，在任何涉及电荷传入和传出半导体晶体的设计中应该将依赖于面生长的能带排列考虑在内，以最大限度地提高其效率。

Ren 等人[107]通过连续电沉积、磁控溅射和二次电沉积合成了夹层 ZnO@Ag@Cu_2O 纳米棒薄膜。在此复合材料中的 Ag 纳米棒发挥了独特的作用。在可见光照射下，活性电子-空穴对通过从 Ag 纳米棒的等离子体偶极子的共振能量转移到 Cu_2O 中产生，并在 ZnO/Cu_2O 界面快速分离。此外，形成的肖特基型异质结可以抑制电子-空穴的复合。

（2）Cu_2O/TiO_2 复合材料

二氧化钛（TiO_2）是一种典型的 n 型半导体，具有优异的光和化学稳定性、强氧化活性，耐腐蚀且无毒。但 TiO_2 的带隙较宽，在可见光辐射下不能有效激活，并且光生电子-空穴对很容易复合[108]。Cu_2O 的导带和价带都位于 TiO_2 的上方[109]。因此，当 TiO_2 与 Cu_2O 耦合时，光生电子从 Cu_2O 的导带转移到 TiO_2 的导带形成 Ti^{4+} 中心，从而延长了光生载流子的寿命。Cu_2O/TiO_2 异质结构的优点受益于 Cu_2O 有吸引力的带隙以及 TiO_2 在水溶液中的高稳定性。

Huang 等人[110]通过醇-水化学沉淀法制备了 Cu_2O/TiO_2 异质结构,并通过在紫外-可见光下降解水中的苏丹红Ⅱ来评估 Cu_2O/TiO_2 异质结构的光催化效率。结果表明,与纯 TiO_2 相比,Cu_2O/TiO_2 异质结构显著提高了光催化活性。增强的活性归因于改善的光吸收和异质结构,这有利于 Cu_2O/TiO_2 异质结构中光引入的电子-空穴对的分离。

Wang 等人[111]通过超声辅助连续化学沉积法在 TiO_2 纳米管阵列上沉积 Cu_2O 纳米颗粒,获得了 Cu_2O/TiO_2 p-n 型异质结光电极。结果表明,Cu_2O/TiO_2 复合光电极在可见光区具有增强的吸收,并且在降解罗丹明 B(RhB)时具有优异的光电催化活性和稳定性(作用机制见图 1.8)。光电流测量表明,Cu_2O/TiO_2 复合材料中光生电子和空穴的分离得到很大改善。创造具有可控晶面的光催化剂已成为增强其活性的重要途径。

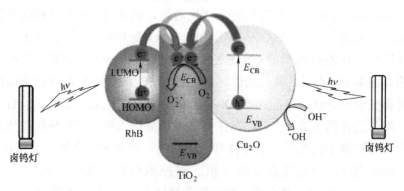

图 1.8 Cu_2O/TiO_2 异质结构降解罗丹明 B 的作用机理[111]

Liu 等人[112]通过简单的水热法制备了 $Cu_2O@TiO_2$ 核壳结构,将锐钛矿型 TiO_2 壳包覆到三种类型的 Cu_2O 晶核上,分别为具有{100}晶面的 Cu_2O 立方体、具有{100}和{111}晶面的 Cu_2O 立方八面体和具有{111}晶面的 Cu_2O 八面体。通过在可见光照射下降解亚甲基蓝和4-硝基苯酚来评价它们的光催化活性。结果表明,在可见光照射下,$Cu_2O@TiO_2$ 八面体由于导带偏移最高,对亚甲基蓝和 4-硝基苯酚的降解表现出最好的光催化性能,其次是 $Cu_2O@TiO_2$ 立方八面体和 $Cu_2O@TiO_2$ 立方体。因此,形成的能带排列控制对于在暴露面上形成异质结的复合光催化剂的设计至关重要。

Liu 等人[113]制备了负载在 TiO_2 纳米片上的 Cu_2O 纳米颗粒,通过一锅水热反应其{001}面暴露出来,在可见光下用于降解苯酚。在 TiO_2 纳米片上负载的 Cu_2O 纳米颗粒导致异质结的形成,有利于光生电子和空穴的有效分离,从而提高光活性。

Yang 等人[114]制备了 Cu_2O/TiO_2 p-n 型异质结催化剂，应用于 4-硝基苯酚的降解，该结构由顶面的 p 型 Cu_2O 纳米线和内壁为 n 型的 TiO_2 纳米颗粒组成。结果表明，Cu_2O/TiO_2 结构比未改性的 TiO_2 纳米管具有更高的降解率。增强的光催化活性可归因于由 Cu_2O 纳米线导致的可见光吸收的扩展以及由在 Cu_2O/TiO_2 p-n 型异质结界面产生的光诱导电位差驱动的光生载流子的有效分离。

Fu 等人[115]介绍了一种双 Z 型 $TiO_2/Ag/Cu_2O$ 光催化系统，通过简单的浸渍煅烧方法将 Cu_2O 负载到 TiO_2 纳米管上，然后通过光沉积方法将 Ag 沉积到光催化剂上。在紫外可见光下，TiO_2 和 Cu_2O 都可以被激发。由于在金属-半导体界面上形成肖特基势垒，TiO_2 导带中的光生电子将转移到 Ag，同时表面等离子共振（SPR）诱导的局部电场将驱动 Ag 的电子与 Cu_2O 价带上的空穴结合。因此，开发了一种用于 $TiO_2/Ag/Cu_2O$ 复合材料光催化反应的双 Z 型电荷转移途径，既可实现高分离效率，又可提高光生电子的氧化还原能力。

（3）Cu_2O/CuO 复合材料

CuO 是一种 p 型半导体，带隙为 1.3~1.6eV[116]。它的导带和价带低于 Cu_2O 的相应带。Jiang 等人[117]报道了通过简便的湿化学方法制备 CuO/Cu_2O 异质体，并且在 Cu_2O 立方体/八面体上可控制备了不同形态的 CuO，即纳米线、四面体（THs）和纳米球（NSs）。通过光催化降解甲基橙作为模型反应，评价了所制备的 CuO/Cu_2O 异质体的光催化活性和稳定性，发现所有制备的 CuO/Cu_2O 的光催化活性和稳定性均显著提高。光催化活性增强的原因主要是形成有利于电荷分离和转移的 II 型能带结构。

如图 1.9 所示，Yu 等人[118]采用无模板水热法制备了 CuO/Cu_2O 复合空心微球，发现合成的 CuO/Cu_2O 复合空心微球在可见光下对甲基橙的光催化脱色表现优于单相 CuO 或 Cu_2O 样品（作用机制见图 1.9）[119]。基于能带理论，CuO 和 Cu_2O 也可以形成异质结构，促进电荷分离，从而提高光催化活性。

Liu 等人[120]通过简便的水热法制备了具有可调成分和形态的 CuO/Cu_2O 空心微球复合材料，并用作高效光催化剂，用于在可见光照射下降解甲基橙的研究。通过改变软模板 Pluronic（P123）的用量、前驱体溶液的 pH 值、水热温度和时间，产物相应地从 CuO/Cu_2O 复合物转变为具有可调节形态的

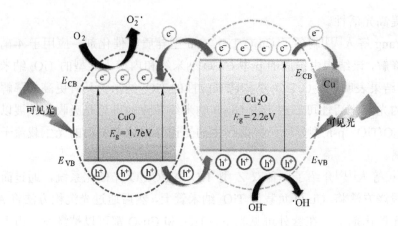

图 1.9 Cu/Cu$_2$O/CuO 异质结构光催化机理[118]

Cu$_2$O 和 Cu/Cu$_2$O 复合物,如大颗粒、多面体、实心和空心微球等。与其他结构相比,CuO/Cu$_2$O 复合空心微球在 300min 内对甲基橙的降解表现出 92.2%的最佳可见光光催化活性,这可以归因于其优异的光捕获能力、独特的空心结构和改进的 CuO/Cu$_2$O 异质结构内的光生载流子分离。

Chen 等人[121]还通过使用简便的软模板辅助水热法制备了 CuO/Cu$_2$O 异质结构复合空心微球。在可见光下降解甲基橙,CuO/Cu$_2$O 表现出比纯 Cu$_2$O 高的可见光光催化活性和重复性,这是由于其高吸附能力、增强的光散射效应、中空结构内的高通量、中空反应区域减少的光生电子和空穴的复合。

Yurddaskal 等人[122]通过在热氧化钛基材上电镀铜涂层制备 Cu$_2$O/CuO 异质结构,并通过在紫外线照射下亚甲基蓝的降解来评价其光催化活性。结果表明,在不同退火温度下样品的微观结构和表面形貌发生了显著变化,500℃退火的样品表现出最高的光催化活性。归因 p 于光催化剂表面的针状 CuO 结构可以增加界面电荷转移并抑制电子-空穴对的复合。Z 型光催化剂的构建已被证实是开发高效可见光复合光催化剂的有效途径之一。

Wang 等人[123]通过简单的两步法设计了一种高效的 Z 型 Cu$_2$O/rGO/CuO 复合光催化剂,包括在 Cu$_2$O 表面接枝 rGO,然后在 Cu$_2$O/rGO 表面原位沉积 CuO 纳米颗粒。结果表明,所有制备的 Cu$_2$O/rGO/CuO 光催化剂都表现出比 Cu$_2$O、Cu$_2$O/rGO 和 Cu$_2$O/CuO 高得多的光催化活性,并且 0.5%(质量分数)Cu$_2$O/rGO/CuO 表现出最高的性能。光催化活性增强归因于 rGO 作为一种新的有效电子载体在 Cu$_2$O/rGO/CuO Z 型体系中有效的电子转移。

Ma 等人[124]使用 CuSO$_4$ 作为单一前驱体,通过简单的碳热还原方法制备了活性炭（AC）负载的铜异质结构复合材料（Cu/Cu$_2$O/CuO/AC），活性炭同时起到还原剂和载体的作用。在复合材料中，尺寸小于 10nm 的 Cu/Cu$_2$O/CuO 复合材料很好地分散在活性炭表面。通过在可见光照射下降解活性艳蓝（KN-R）来测试催化剂的光催化活性。结果表明，与其他催化剂相比，Cu/Cu$_2$O/CuO/活性炭异质结构复合材料表现出优异的光催化活性，活性炭和光活性铜物种之间的协同作用，以及界面结构的存在，如 Cu$_2$O/CuO 异质结、Cu/Cu$_2$O（或 CuO）肖特基异质结和 Cu$_2$O/Cu/CuO 异质结，都是其优异性能的原因。

Li 等人[125]开发了一种新型的自立式柔性可见光光催化剂 Cu/Cu$_2$O/CuO 异质结网，该催化剂反应后可以通过使用低成本的 Cu 网作为基质在空气中一步煅烧即可回收。由于 Cu$_2$O 的导带和价带都位于 CuO 的下方，在可见光照射下，Cu$_2$O 的光激发电子转移到 CuO 的导带，而 CuO 的光激发空穴转移到 Cu$_2$O 的价带。此外，Cu 骨架还充当良好的电子受体，接受来自 Cu$_2$O 导带的光激发电子，因此，Cu/Cu$_2$O/CuO 异质结的混合窄带隙确保其在可见光区的宽吸收带和有效的电子-空穴分离，因此在可见光照射下具有显著的光催化活性。

Ajmal 等人[126]制备了非贵金属负载的 Cu$_2$O/CuO/TiO$_2$ 催化剂并用于纺织染料的光降解，并发现使用 4%（质量分数）Cu$_2$O/CuO/TiO$_2$ 可以有效地实现对 RB49 的光催化降解。Cu$_2$O/CuO/TiO$_2$ 中 Cu 的直接电子俘获特性，提高了催化剂的染料降解能力。TEM 和 X 射线光电子能谱（XPS）结果证实了氧化物 Cu$_2$O/CuO 中存在 Cu。

（4）Cu$_2$O/Fe$_2$O$_3$ 复合材料

Fe$_2$O$_3$ 由于其窄带隙、合适的带边电位、地球上含量丰富且无毒而在各种应用中得到了广泛的关注[127]。Lakhera 等人[128]通过一步简便的水热法制备了 α-Fe$_2$O$_3$/Cu$_2$O 混合氧化物光催化剂。α-Fe$_2$O$_3$/Cu$_2$O 光催化剂的光致发光强度降低意味着 α-Fe$_2$O$_3$ 在 Cu$_2$O 上的负载增强了界面处的电荷载流子分离和转移。5%（质量分数）α-Fe$_2$O$_3$ 负载的 Cu$_2$O 表现出最高的光降解活性，与纯 Cu$_2$O 和 α-Fe$_2$O$_3$ 相比，甲基橙（MO）的光降解率分别提高了近 30%和 95%。α-Fe$_2$O$_3$/Cu$_2$O 光催化剂优异的光降解活性主要归功于可见光吸收的增强、有效的载流子分离和转移（见图 1.10）。

Tian 等人[129]开发了一种独特的三元单核双壳异质结构，由具有管状形

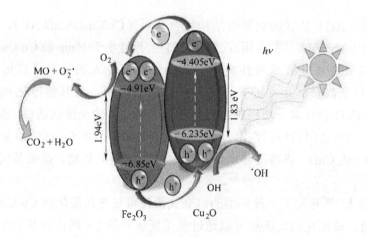

图 1.10　α-Fe_2O_3/Cu_2O 异质结构光催化降解甲基橙作用机理[120]

态的 α-Fe_2O_3@SnO_2@Cu_2O 组成。在制备过程中，使用阴离子辅助水热法沉积 α-Fe_2O_3，使用种子介导沉积策略 SnO_2，最后使用老化工艺沉积 Cu_2O 层以完成管状三元异质结构。在模拟阳光照射下这些三元异质结构对罗丹明 B（RhB）的光降解表现出优异的光催化性能。光催化活性的增强是由于通过 p-n 型异质结促进了光生电子-空穴对的有效电荷分离。

Li 等人[130]首次通过简便的电化学工艺合成了一种新型 Z 型 Cu_2O/GO/α-Fe_2O_3 纳米管阵列（Cu_2O/GO/FNA）复合材料。通过在可见光照射下降解亚甲基蓝来评价其光电化学性能，结果表明，与单组分或双组分体系相比，由于有效的光生电子空穴，Cu_2O/GO/FNA 复合材料表现出良好的光活性和光稳定性。

Shen 等人[131]合成了 rGO/Cu_2O/Fe_2O_3 复合材料的全固态 Z 型系统，并用于同时制氢和四环素降解以测试其光催化性能。所得 rGO/Cu_2O/Fe_2O_3 复合材料在产氢和四环素降解方面表现出比单一 Cu_2O 或 Fe_2O_3 更优异的光催化性能。这是因为还原氧化石墨烯（rGO）作为固态介质可以有效地将光生电子从 Fe_2O_3 的导带传输到 Cu_2O 的价带，这在产氢和四环素降解过程中起关键作用。

（5）其他金属氧化物/Cu_2O 复合材料

除了上面讨论的 Cu_2O/金属氧化物复合材料外，由 Cu_2O 和其他金属氧化物组成的多种混合金属氧化物，也被用来有效利用太阳光进行各种应用。例如，Luo 等人[132]通过两步法制备了 Cu_2O 改性的 Bi_2O_3 纳米球。在复合材料中，Cu_2O 分散在 Bi_2O_3 纳米球的表面，所有 Cu_2O 改性的 Bi_2O_3 纳米球均呈现出大小为 80~150nm 的均匀纳米球。与 Cu_2O 和 Bi_2O_3 相比，对于罗丹明 B 的降解，复合材料表现出增强的光催化活性。优异的性能归因于更高的 BET（比表面积）、带隙变窄以及 Cu_2O 改性的 Bi_2O_3 纳米球中的界面电荷转移效应。

Shen 等人[133]通过原位沉淀法在 rGO 表面沉积 Cu_2O 和 Bi_2O_3 纳米晶体，制备了 rGO/Cu_2O/Bi_2O_3 复合光催化剂。在复合材料中，Cu_2O 和 Bi_2O_3 纳米晶体均匀且独立地分布在 rGO 上。对于可见光照射下四环素（TC）的光降解，三组分 rGO/Cu_2O/Bi_2O_3 复合材料的光催化性能得到明显提高。这是由于构建了 Z 型光催化系统，其中 rGO 作为固态介质，用于将光生电子从 Bi_2O_3 的导带转移到 Cu_2O 的价带。

Wei 等人[134]在钛衬底上用连续的阴极电极定位法制备了 WO_3/p-Cu_2O 复合膜。在模拟自然光照射下，通过 OrangeⅡ的分解来评价光催化活性。结果表明，与单独使用 WO_3 和 Cu_2O 相比，WO_3/p-Cu_2O 表现出更高的光催化活性。这是因为在这个 p-n 型异质结中，WO_3 中的电子和 p 型 Cu_2O 中的空穴将通过两个半导体之间的界面转移而结合，但各自半导体中光生电荷的复合将被抑制，从而产生高光催化活性。

Li 等人[135]通过快速简单的途径大量制备了磁性可回收的 Cu_2O/Fe_3O_4 复合光催化剂，并通过在可见光照射下使用甲基橙降解反应测试其光催化性能。光催化剂在 90min 内达到 90%的降解率，Cu_2O/Fe_3O_4 的最佳重量比为 4∶6，连续使用 5 次后，其初始活性仍然保持在 95%以上，表明具有较高的可重复使用性。

1.3.3　Cu_2O/其他化合物

（1）Cu_2O/Ag 基化合物

Bi 等人[136]首次报道了 Ag_3PO_4 在水氧化和有机污染物处理领域具有出色的可见光驱动光催化性能，这种新型半导体因其合适的带隙位置、安全性和高光催化能力而备受关注。Li 等人[137]通过化学方法在低温下合成了一种新型 Z 型八面体 Cu_2O/Ag_3PO_4 复合材料（见图 1.11）。在可见光下其光催化性能通过亚甲基蓝降解来评价。反应 8min 后，90%的亚甲基蓝被 Cu_2O/Ag_3PO_4 降解，然而，Ag_3PO_4 和 Cu_2O 的光降解率分别为 63%和 24%。

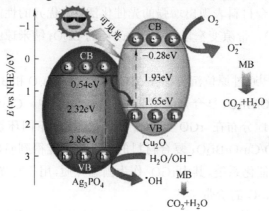

图 1.11　Cu_2O/Ag_3PO_4 异质结构光催化机理[137]

Hou 等人[138]通过液相还原和化学沉积方法制造了一种新型的 Ag_3PO_4@Cu_2O 核壳异质结光催化剂。在复合材料中，具有花状结构的 Cu_2O 负载在 Ag_3PO 纳米棒的表面。Ag_3PO_4@Cu_2O 复合材料表现出最高的催化活性，在照射 20min 后近 97%的亚甲基蓝被降解。

AgBr 是一种带隙为 2.6eV 的 n 型半导体[139]，在光催化领域得到了广泛的研究并表现出出色的光催化活性。为了克服 Cu_2O 的缺点，使用少量 AgBr/Ag 对 Cu_2O 表面进行改性。Liu 等人[78]通过简单的原位光还原方法将 Ag@AgBr 负载在八面体 Cu_2O 基底的{111}制备了一种新的复合光催化剂 Ag@AgBr/Cu_2O。在复合材料中，Ag@AgBr 纳米粒子很好地分散在 Cu_2O 纳米粒子上，具有窄的尺寸分布和可控的尺寸（10~30nm）。通过在可见光照射下降解亚甲基蓝来评价其光催化活性。与纯 Cu_2O 相比，Ag@AgBr/Cu_2O 复合物表现出更强的可见光吸收能力和更高的光催化活性，15%（质量分数）Ag@AgBr/Cu_2O 样品表现出最好的光催化活性，光照 90min 后亚甲基蓝的降解率为 93.28%。

Hu 等人[140]通过沉积-沉淀法制备了一种负载在介孔氧化铝（Cu_2O/Ag/AgBr/Al_2O_3）上的等离子体光催化剂 Cu_2O/Ag/AgBr。Cu_2O 与 Ag 纳米颗粒和 AgBr 的耦合加速了界面电子转移过程，使溶解的 Ag^+ 快速光还原，进而提高了 Cu_2O 的稳定性。因此，对于有毒持久性有机污染物的降解，该催化剂在可见光照射下表现出较高的光催化活性和稳定性。

He 等人[141]通过氧化还原过程和光辅助沉积制备了 $Cu_2O/Cu/AgBr/Ag$ 光催化剂。在这种复合材料中，Cu NPs 可以在 Cu_2O 和 AgBr 之间可控地生长。可见光照射下 50min，在 Cu_2O 和 AgBr 之间没有 Cu 纳米颗粒的情况下，仅观察到甲基橙较低的光催化降解（51%）。然而，在 Cu 纳米颗粒的存在下，约 98%的甲基橙可以被降解（见图 1.12）。这是因为引入 Cu 纳米颗粒有助于加速 Cu_2O 和 AgBr 之间界面处的载流子转移，并且为此过程提出了 Z 型异质结机理。

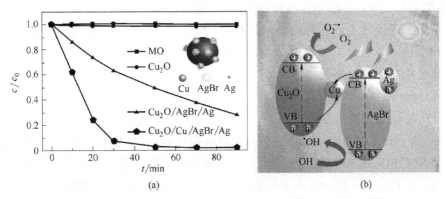

图 1.12　$Cu_2O/Cu/AgBr/Ag$ 异质结光催化降解甲基橙[141]

(a) 光催化降解结果比较；(b) 光催化机理

在各种宽带隙半导体光催化剂中，AgCl 是一种 n 型半导体，其直接带隙和间接带隙分别为 5.15eV 和 3.25eV[142]。此外，由于其三维电子轨道 d_{10}-d_{10} 间的相互作用和更短的 Ag/Ag 键，AgCl 具有更小的电子有效质量从而导致更高的电子迁移率和更长的光生载流子寿命[143,144]。因此，AgCl 与 Ag 纳米粒子偶联具有良好的光催化性能，这主要归因于 Ag 纳米粒子的表面等离子体效应和有效的电荷转移[145-147]。为了克服 Cu_2O 电荷复合的缺点，还使用 AgCl/Ag 对 Cu_2O 进行改性。Lou 等人[148]在室温下使用 Cu_2O/Ag 微立方体作为模板，$CuCl_2$ 作为氧化剂，通过一种简便的方法制备了分层 Cu_2O/Ag/AgCl 微立方体。通过 MO 的降解来评价其光催化性能，在最佳条件下，16min 内可降解 93%的 MO。这是因为 Cu_2O 阻止了 AgCl 的电子重新结合并促进光生载流子的分离，从而提高了光催化性能。

（2）Cu_2O/Bi 基化合物

近年来，铋基光催化剂受到了广泛关注[149]。由于杂化的 O 2p 轨道和 Bi^{3+}

的 $6s^2$ 轨道的杂化，许多含 Bi^{3+} 的化合物具有较窄的禁带并表现出高的可见光光催化活性。

具有单斜结构的 $BiVO_4$ 是一种 n 型半导体，其离散带隙约为 2.3eV[150]，已被广泛应用于光催化领域。Cu_2O 中的导带和价带电位都比 $BiVO_4$ 中的更负。这些半导体能带的相对位置表明它们之间有可能形成 p-n 型异质结。在最近的研究中，p-n 型异质结 $Cu_2O/BiVO_4$ 复合光催化剂可以通过不同的方法合成[151,152]。例如，Wang 等人[153]通过将水热工艺与多元醇相结合，制备了 p-n 型异质结 $Cu_2O/BiVO_4$ 异质纳米结构。通过在可见光照射下降解亚甲基蓝和苯酚来评价其光催化性能。p 型 Cu_2O 和 n 型 $BiVO_4$ 界面之间形成的 p-n 型异质结可以有效地分离光生电子和空穴，并且可以显著减少电子-空穴对的复合从而提高光催化活性。

Deng 等人[154]通过简单的 Cu_2O 颗粒浸渍法和光还原法制备了双 Z 型 $Cu_2O/Ag/BiVO_4$ 纳米复合材料，并在可见光照射下用于降解四环素。图 1.13（a）为 p-n 型异质结 $Cu_2O/BiVO_4$ 光降解四环素的机理图，图 1.13（b）为双 Z 型 $Cu_2O/Ag/BiVO_4$ 纳米复合材料光催化降解四环素的机理图。活性实验表明，在 Cu_2O 和 2% Ag 的最佳涂层含量下，三元 $Cu_2O/Ag/BiVO_4$ 纳米复合材料的四环素去除效率可达 91.22%，高于纯 $BiVO_4$（42.9%）和二元 $Cu_2O/BiVO_4$（65.17%）、$BiVO_4/Ag$（72.63%）纳米复合材料。$Cu_2O/Ag/BiVO_4$ 光催化活性的增强归因于 Cu_2O、Ag 和 $BiVO_4$ 的协同作用，特别是表面等离子体共振效应和金属银带来的已建立的局部电场。

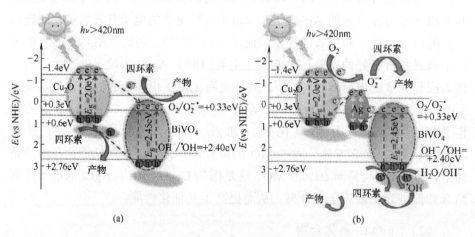

图 1.13 $Cu_2O/BiVO_4$ 异质结构（a）和 $Cu_2O/Ag/BiVO_4$ 异质结构（b）光催化机理[155]

根据文献[155,156]，三维 Bi_2WO_6 在可见光激发下，电荷可以从其 O2p Bi6s 杂化轨道转移到空的 W 5d 轨道。此外，Bi_2WO_6 独特的花状结构具有较大的比表面积，可以提供大量的反应位点，并且光散射效应可有效提高可见光的利用率[14-16]。如图 1.14 所示，Liu 等人[157]通过界面自组装方法制备了装饰花状 Bi_2WO_6 的 Cu_2O 纳米点。在复合材料中，平均直径为 20nm 的 Cu_2O 纳米点均匀地分散在 Bi_2WO_6 的表面，3%（质量分数）Cu_2O/Bi_2WO_6 复合材料对亚甲基蓝的降解率最高，分别是纯 Bi_2WO_6 和 Cu_2O 的 2.14 倍和 12.25 倍。这种增强的光催化活性归因于可见光吸收效率的增强以及接触界面的强相互作用所致的有效光生电荷分离。

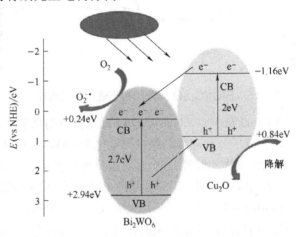

图 1.14　Cu_2O/Bi_2WO_6 异质结构光催化作用机理[158]

Li 等人[158]制备了一种新型 Z 型 $Cu_2O/Au/BiPO_4$ 复合材料，并用于甲基橙降解。在光催化反应中，金属 Au 不仅可以作为固态电子介体，还能吸收入射光中的光子，并在该杂化体系中表现出表面等离子体共振效应。这种 Z 型异质结可以提高光生载流子的分离效率，有利于抑制电子-空穴的复合。因此，该复合材料表现出优异的性能，在照射 60min 后甲基橙被完全降解。

如图 1.15 所示，Xia 等人[159]首次通过低温耦合还原法设计合成了新型 $BiOI/Cu_2O$ 复合材料。这种 $BiOI/Cu_2O$ 复合材料呈现三维分层菜花状结构，由 BiOI 纳米片和 Cu_2O 立方亚微米结构组成。在 $BiOI/Cu_2O$ 复合材料中，BiOI 和 Cu_2O 在界面处的协同作用可以大大增强光生电子-空穴对的分离，提高光生电荷载流子的迁移效率并有效防止光生电子和空穴的复合，从而产生优异的光催化性能。

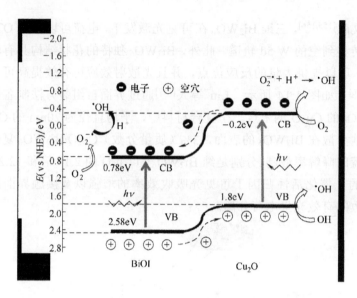

图 1.15 BiOI/Cu$_2$O 异质结构光催化作用机理[159]

Cui 等人[160]将 Cu$_2$O 量子点（QDs-Cu$_2$O）通过简单的化学法结合到三维花状分级 BiOBr 中，合成了一种高效的新型光催化剂 QDs-Cu$_2$O/BiOBr，并在可见光下用于降解有机污染物的研究。3%（质量分数）QDs-Cu$_2$O/BiOBr 复合材料对苯酚和亚甲基蓝的降解率最高，分别是纯 BiOBr 的 11.8 倍和 1.4 倍。通过将 Cu$_2$O QDs 与 BiOBr 结合，增强了可见光吸收效率，另外，由于其界面的紧密接触及良好的能带结构排列使光产生的载流子能够进行有效分离。

（3）g-C$_3$N$_4$/Cu$_2$O 复合材料

石墨氮化碳将（g-C$_3$N$_4$）是一种新型的无金属聚合物材料，其带隙较小，为 2.7eV，可以吸收更多的自然阳光。g-C$_3$N$_4$ 对热化学和光化学侵蚀非常稳定，是一种理想的光催化材料。g-C$_3$N$_4$ 和 Cu$_2$O 半导体结合形成异质结是促进电荷分离、减少它们复合的有效途径。

Tian 等人[161]将 Cu$_2$O 沉积到 g-C$_3$N$_4$ 表面，通过水热法制备了 g-C$_3$N$_4$/Cu$_2$O 的 p-n 型异质结。光催化测试表明，在可见光照射下 g-C$_3$N$_4$-Cu$_2$O 异质结具有优异的光催化降解 MO 的活性，远高于纯 g-C$_3$N$_4$ 或 Cu$_2$O。这归因于异质结界面处光生电子-空穴对的高度分离和快速转移。

如图 1.16 所示，Li 等人[162]还在谷氨酸存在下通过水热合成和高温煅烧制备了具有 p-n 型异质结结构的 g-C$_3$N$_4$/Cu$_2$O。通过添加谷氨酸，复合材料的比表面积和反应位点均增加。此外，由于异质结中电荷重组率降低，且接触界面更紧密，使可见光利用效率提高。

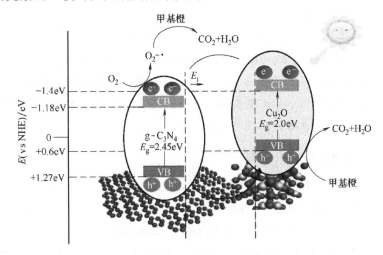

图1.16　g-C_3N_4/Cu_2O异质结构光催化作用机理[162]

Liu等人[163]通过结合溶剂热和化学吸附的方法制备了Cu_2O@g-C_3N_4核@壳复合材料，其中g-C_3N_4通过自发吸附过程成功地涂覆到Cu_2O纳米球上。结果表明，g-C_3N_4包覆的Cu_2O纳米球具有显著提高的光催化制氢活性，这归因于g-C_3N_4的接触面积大、特定能带结构以及Cu_2O和g-C_3N_4界面处电荷分离增强的协同作用。

Yan等人[164]通过简便的自组装方法制造了一种高效的Cu_2O/g-C_3N_4/rGO气凝胶光催化剂。负载在还原氧化石墨烯片上的g-C_3N_4/Cu_2O异质结导致可见光范围内的吸收增加并提高其光降解活性，并且三元复合气凝胶在80min内表现出96%的亚甲基蓝和83%MO的降解效率。

Liang等人[165]通过在多孔g-C_3N_4上化学吸附Ag掺杂的Cu_2O合成了多孔g-C_3N_4/Ag/Cu_2O复合材料，并在可见光照射下用于降解罗丹明B。当Ag/Cu_2O的含量（质量分数）为15%时，在可见光照射50min后85%的罗丹明B被降解。光催化性能增强归因于增强了对可见光的利用，增加了界面处光生电子-空穴对的分离和快速转移，以及通过向复合材料引入多孔结构来增大比表面积。

Zuo等人[166]通过一锅氧化还原法制备了高性能光催化剂，凹凸棒石/Cu_2O/Cu/g-C_3N_4。在制备过程中，棒状凹凸棒石表面的超细CuO纳米颗粒在三聚氰胺热缩过程中被NH_3气体原位还原生成Cu和少数Cu_2O。同时，将生成的g-C_3N_4膜均匀包裹在凹凸棒石/Cu_2O/Cu表面，组装Z型Cu_2O/Cu/

g-C$_3$N$_4$ 异质结构。所得的 ATP/Cu$_2$O/Cu/g-C$_3$N$_4$ 催化剂对氯霉素的降解表现出优异的光催化性能。

Kumar 等人[74]通过简便的共沉淀方法制备了一种新型磁性四元 BiOCl/g-C$_3$N$_4$/Cu$_2$O/Fe$_3$O$_4$（BGC-F）纳米异质结。BGC-F 表现出优异的光催化活性，在可见光下照 60min 后，100μmol/L 磺胺甲噁唑被降解 99.5%，这是由于形成了有效的 p-n-p 型异质结（BiOCl/g-C$_3$N$_4$/Cu$_2$O）导致能带移动、内置电场的形成和电荷分离效率的提高。

（4）CdS/Cu$_2$O 复合材料

CdS 是一种 n 型半导体，带隙约为 2.4eV，在光催化[167]、太阳能电池[168]、生物标记[169]和环境传感器[170]领域有广泛的应用。如图 1.17 所示，Liu 等人[171]通过溶剂-热工艺和化学浴沉积工艺合成了 CdS/Cu$_2$O 异质结构材料。在复合材料中，CdS 纳米线（NWs）的表面用球形 Cu$_2$O 装饰，其直径范围为 100~200nm。通过光降解 MO 来研究其光催化活性，CdS/Cu$_2$O 复合材料在照射 120min 后达到约 96%的降解率，在可见光下，Cu$_2$O 导带（CB）中的光生电子有向 CdS 的 CB 转移的趋势，而光生空穴则从 CdS 的价带（VB）向 Cu$_2$O 的价带方向转移，光生电子和空穴进而有效分离。在光催化降解过程中，CdS/Cu$_2$O 复合材料的特殊异质结构起着重要作用。

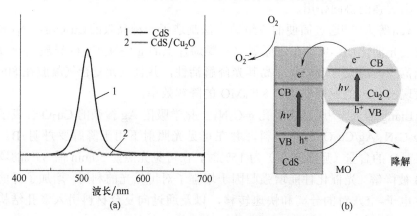

图 1.17　CdS/Cu$_2$O 异质结构[171]

(a) 光催化剂的紫外可见光谱；(b) 光催化机理

（5）Cu$_2$O/ZnWO$_4$ 复合材料

最近，ZnWO$_4$ 因其在紫外光下降解有机污染物的优异性能而受到广泛关注[172]。同时，ZnWO$_4$ 光催化剂还具有化学性质稳定、易于开发等优点。由于 ZnWO$_4$ 和 Cu$_2$O 的能级结构相匹配使得两者在界面处紧密结合，这有利于加速光生电荷载流子的分离，从而提高光催化活性。

如图 1.18 所示，Tian 等人[173]通过沉淀法用 Cu$_2$O 纳米颗粒装饰 ZnWO$_4$ 纳米棒的表面（Cu$_2$O/ZnWO$_4$）。在复合材料中，Cu$_2$O 纳米颗粒紧密沉积在 ZnWO$_4$ 表面，平均直径为 20nm。可见光下对 MB 进行降解，质量分数为 5% Cu$_2$O/ZnWO$_4$ 光催化剂显示出最高的降解效率，分别比纯 ZnWO$_4$ 和 Cu$_2$O 高 7.8 倍和 2 倍。这是因为纳米颗粒不仅促进了可见光的吸收，而且促进了光生载流子的分离。

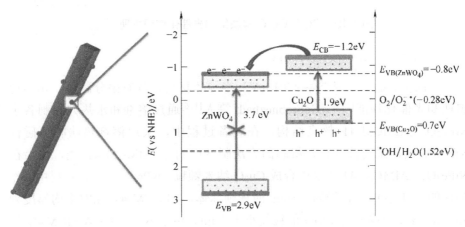

图 1.18 Cu$_2$O/ZnWO$_4$ 异质结构[173]

（6）Cu$_2$S/Cu$_2$O 复合材料

硫化亚铜（Cu$_2$S）是一种带隙为 1.20eV 的半导体。Cu$_2$O 的导带（CB）比 Cu$_2$S 的导带更正，Cu$_2$O 导带中的光生电子向 Cu$_2$S 的导带转移，而光生空穴则以相反的方向转移。如图 1.19 所示，Yue 等人[174]通过简便的共沉淀和煅烧方法合成了 Cu$_2$S 修饰的 Cu$_2$O 纳米复合材料，并用于去除有机污染物，包括刚果红、甲基橙（MO）和四环素（TC）。与之前报道的研究相比，Cu$_2$O/Cu$_2$S 纳米复合材料（O/S 摩尔比为 9∶1）具有更优越的光催化性能，可在 120min 内去除约 99.8% 的刚果红、90.1% 甲基橙和 84.8% 四环素。吸附能力的增强主要归因于纳米复合材料与有机物之间的静电相互作用和络合。此外，适量的 Cu$_2$S 不仅可以增加纳米复合材料的比表面积，还可以有效增强可见光吸收，抑制光生电子和空穴的复合。

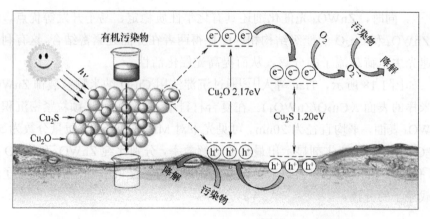

图 1.19 Cu$_2$S/Cu$_2$O 异质结构光催化作用机理[174]

（7）NiFe$_2$O$_4$/Cu$_2$O 复合材料

NiFe$_2$O$_4$ 是一种窄带隙（2.19eV）半导体材料，具有磁分离性、化学稳定性和光催化性能[175]。Sohrabnezhad 等人[176]通过溶剂和水热方法制备了 NiFe$_2$O$_4$@AlMCM-41 核壳结构。在制备过程中，采用溶剂热法制备磁性 NiFe$_2$O$_4$ 核，采用液晶模板机制的方法制备中间体 AlMCM-41 壳，随后在 NiFe$_2$O$_4$@AlMCM-41 核壳中合成 Cu$_2$O 纳米颗粒（NPs）。在复合材料中，作为核的 NiFe$_2$O$_4$ 是直径为 68nm 的规则球体，AlMCM-41 壳的平均厚度为 35nm，核壳中掺入的 Cu$_2$O 的粒径小于 5nm。由于 NiFe$_2$O$_4$@AlMCM-41 核壳减少了 Cu$_2$O 中电子-空穴对的复合，该复合材料表现出优异的催化性能，在可见光下照射 60min 后亚甲基蓝去除率高达 90%。He 等人[177]通过两步法制备了磁性可分离的 NiFe$_2$O$_4$/Cu$_2$O 复合材料，并用于模拟太阳光照射下的 MB 降解。NiFe$_2$O$_4$/Cu$_2$O 复合材料表现出比纯 Cu$_2$O 和 NiFe$_2$O$_4$ 更高的光催化活性，在模拟太阳光照射 60min 下，NiFe$_2$O$_4$/Cu$_2$O 复合材料的光催化效率接近 98%。这归因于 NiFe$_2$O$_4$ 和 Cu$_2$O 的共催化作用，导致光生电子-空穴对的有效分离和光生电子从 NiFe$_2$O$_4$ 到 Cu$_2$O 的高效转移。

（8）SrTiO$_3$/Cu$_2$O 复合材料

SrTiO$_3$ 是一种重要的 n 型半导体，具有立方钙钛矿结构，带隙约为 3.2eV[178]，因其具有热稳定性、良好的耐热性、耐腐蚀性等而引起广泛关注[179,180]。当 Cu$_2$O 与 SrTiO$_3$ 结合时，可以获得 p-n 型异质结 SrTiO$_3$/Cu$_2$O 复合材料。Xia 等人[181]首次通过两步法制备了 p-n 型异质结 SrTiO$_3$/Cu$_2$O 复合材料，并用于污染物 MB 的降解。所得 SrTiO$_3$/Cu$_2$O 复合材料的光催化活性远高于纯 Cu$_2$O 和 SrTiO$_3$，在模拟太阳光照射 60min 下催化效率接近 90.4%，这归因于 SrTiO$_3$ 和 Cu$_2$O 之间形成 p-n 型异质结的协同作用。

1.4 结语

如上所述，Cu$_2$O 是一种 p 型半导体，其带隙为 2.0~2.2eV。与传统的光催化材料（TiO$_2$ 和 ZnO）相比，Cu$_2$O 的带隙要窄得多，可以在可见光下发生光催化反应。研究者通过其与贵金属、金属氧化物和负载材料结合形成半导体异质结、肖特基型异质结或 Z 型异质结，提高了光利用率，降低了光生电子-空穴对复合率，从而提高了 Cu$_2$O 的光催化性能。如上所述，抑制光生电子-空穴对的复合是提高 Cu$_2$O 基光催化剂催化性能的关键因素。研究者已经考虑了多种有效的策略来提高其性能。例如，负载在大比表面积的载体上以增加活性位点和提高吸附性，与导电材料耦合以有效分离 Cu$_2$O 中的电子，与其他半导体复合形成异质结以减少光生电子-空穴对的复合。

尽管已经报道了 Cu$_2$O 基光催化剂的许多研究，但基于 Cu$_2$O 基光催化剂的活性和稳定性仍然存在一些严重的问题。因此，下面将通过一些具体的研究实例，如将 Cu$_2$O 与其他半导体石墨氮化碳（g-C$_3$N$_4$）、金属有机骨架材料（UiO-66-NH$_2$）、碳化钛（Ti$_3$C$_2$Tx）形成异质结或者负载在大比表面积和导电材料超交联有机微孔聚合物（KAPs-B）、氮掺杂有序介孔碳（N-CMK-3）、膨胀石墨（EG）、碳纳米管（CNTs）上来实现光生电子-空穴对在催化剂中的快速分离，从而开发出高效且稳定的 Cu$_2$O 基光催化剂[182-186]。此外，还将讨论降解过程的光催化反应机制，有望获得用于降解有机污染物的新型 Cu$_2$O 基光催化剂。

参考文献

[1] Meng X, Li Z, Chen J, et al. Enhanced visible light-induced photocatalytic activity of surface-modified BiOBr with Pd nanoparticles [J]. Appl Surf Sci, 2018, 433:76-87.

[2] Yu Z, Li F, Yang Q. Nature-mimic method to fabricate polydopamine/graphitic carbon

nitride for enhancing photocatalytic degradation performance [J]. ACS Sustain Chem Eng, 2017, 5: 7840-7850.

[3] Yoon S, Kim M, Kim I S, et al. Manipulation of cuprous oxide surfaces for improving their photocatalytic activity [J]. J Mater Chem A, 2014, 2: 11621-11627.

[4] Wei Z, Xin T, Xiao W D, et al. Novel pn heterojunction photocatalyst fabricated by flower-like $BiVO_4$ and Ag_2S nanoparticles: simple synthesis and excellent photocatalytic performance [J]. Chem Eng J, 2019, 361: 1173-1181.

[5] Zhong S, Lv C, Shen M, et al. Synthesis of Bi_2WO_6 photocatalyst modified by SD-BS and photocatalytic performance under visible light [J]. J Mater Sci-Mater EL, 2019, 30: 4152-4163.

[6] Dadigala R, Bandi B R. Gangapuram V, Construction of in situ self-assembled $FeWO_4$/g-C_3N_4 nanosheet heterostructured Z-scheme photocatalysts for enhanced photocatalytic degradation of rhodamine B and tetracycline [J]. Nano Adv, 2019, 1: 322-333.

[7] Li R, Zhang F, Wang D, et al. Spatial separation of photogenerated electrons and holes among {010} and {110} crystal facets of $BiVO_4$ [J]. Nat Commun, 2013, 4: 1-7.

[8] Zhang J, Ma H, Liu Z. Highly efficient photocatalyst based on all oxides WO_3/Cu_2O heterojunction for photoelectrochemical water splitting [J]. Appl Catal B, 2017, 201: 84-91.

[9] Cai T, Wang L, Liu Y, et al. Ag_3PO_4/Ti_3C_2 MXene interface materials as a Schottky catalyst with enhanced photocatalytic activities and antiphotocorrosion performance [J]. Appl Catal B, 2018, 239: 545-554.

[10] Fang H, Pan Y, Yin M, et al. Facile synthesis of ternary Ti_3C_2-OH/In_2S_3/CdS composite with efficient adsorption and photocatalytic performance towards organic dyes [J]. J Solid State Chem, 2019, 280: 120981.

[11] Ng C H B, Fan W Y, Shape evolution of Cu_2O nanostructures via kinetic and thermodynamic controlled growth [J]. J Phys Chem B, 2006, 110: 20801-20807.

[12] Zhang J T, Liu J F, Peng Q, et al. Nearly monodisperse Cu_2O and CuO nanospheres: preparation and applications for sensitive gas sensors [J]. Chem Mater, 2006, 18: 867-871.

[13] Zhang H, Zhu Q, Zhang Y, et al. One-pot synthesis and hierarchical assembly of hollow Cu_2O microspheres with nanocrystals-composed porous multishell and their gas-sensing properties [J]. Adv Funct Mater, 2007, 17: 2766-2771.

[14] Zhong J H, Li G R, Wang Z L, et al. Facile electrochemical synthesis of hexagonal Cu_2O nanotube arrays and their application [J]. Inorg Chem, 2011, 50: 757-763.

[15] Kuo C H, Yang YC, Gwo S, et al. Facet-dependent and Au nanocrystal-enhanced electrical and photocatalytic properties of Au-Cu_2O Core-shell heterostructures [J]. J Am Chem Soc, 2011, 133:1052.

[16] Xiang J Y, Wang X L, Xia X H, et al. Enhanced high rate properties of ordered porous Cu_2O film as anode for lithium ion batteries [J]. Electrochim Acta, 2010, 55: 4921-4925.

[17] White B, Yin M, Hall A, et al. Complete CO oxidation over Cu_2O nanoparticles supported on silica gel [J]. Nano Lett, 2006, 6: 2095-2098.

[18] Leng M, Liu M, Zhang Y, et al, Polyhedral 50-facet Cu_2O microcrystals partially enclosed by {311} high-index planes: synthesis and enhanced catalytic CO oxidation activity [J]. J Am Chem Soc, 2010, 132:17084-17087.

[19] Paracchino A, Laporte V, Sivula K, et al. Highly active oxide photocathode for photoelectrochemical water reduction [J]. Nat Mater, 2011, 10: 456-461.

[20] Kondo J N. Cu_2O as a photocatalyst for overall water splitting under visible irradiation [J]. Chem Commun, 1998, 3:357-358.

[21] Roos A, Karlsson B. Properties of oxidized copper surfaces for solar applications [J]. Sol Energ Mater, 1983, 7: 467-480.

[22] Bao H, Zhang W, Shang D, et al. Shape-dependent reducibility of cuprous oxide nanocrystals [J]. J Phys Chem C, 2010, 114: 6676-6680.

[23] Kuo C H, Chen C H, Huang M H. Seed-mediated synthesis of monodispersed Cu_2O nanocubes with five different size ranges from 40 to 420 nm [J]. Adv Funct Mater, 2007, 17: 3773-3780.

[24] Xu Y, Wang H, Yu Y, et al. Cu_2O nanocrystals: surfactant-free room-temperature morphology-modulated synthesis and shape-dependent heterogeneous organic catalytic activities [J]. J Phys Chem C, 2011, 15: 15288-15296.

[25] Zhang Y, Deng B, Zhang T, et al. Shape effects of Cu_2O polyhedral microcrystals on photocatalytic activity [J]. J Phys Chem C, 2010, 114: 5073-5079.

[26] Zhang L, Wang H. Interior structural tailoring of Cu_2O shell-in-shell nanostructures through multistep Ostwald ripening [J]. J Phys Chem C, 2011, 115: 18479-18485.

[27] Lu C, Qi L, Yang J, et al. One-pot synthesis of octahedral Cu_2O nanocages via a

catalytic solution route [J]. Adv Mater, 2005, 17: 2562-2567.

[28] Kuo C H, Huang M H. Fabrication of truncated rhombic dodecahedral Cu_2O nanocages and nanoframes by particle aggregation and acidic etching [J]. J Am Chem Soc, 2008, 13: 12815-12820.

[29] Orel Z C, Anžlovar A, Dražić G, et al. Cuprous oxide nanowires prepared by an additive-free polyol process [J]. Cryst Growth Des, 2007, 7: 453-458.

[30] Grez P, Herrera F, Riveros G, et al. Synthesis and characterization of p-Cu_2O nanowires arrays [J]. Mater Lett, 2013, 92:413-416.

[31] Zhang Z, Che H, Wang Y, et al. Facile synthesis of mesoporous Cu_2O microspheres with improved catalytic property for dimethyldichlorosilane synthesis [J]. Ind Eng Chem Res, 2012, 51: 1264-1274.

[32] Ko E, Choi J, Okamoto K, et al. Cu_2O nanowires in an alumina template: Electrochemical conditions for the synthesis and photoluminescence characteristics [J]. ChemPhysChem, 2006, 7: 1505-1509.

[33] Gou L, Murphy C J. Solution-phase synthesis of Cu_2O nanocubes [J]. Nano Lett, 2003,3: 231-234.

[34] Kim M H, Lim B, Lee E P, et al. Polyol synthesis of Cu_2O nanoparticles: Use of chloride to promote the formation of a cubic morphology[J]. Mater Chem, 2008, 18: 4069-4073.

[35] Kuo C H, Huang M H. Morphologically controlled synthesis of Cu_2O nanocrystals and their properties [J]. Nano Today, 2010, 5: 106-116.

[36] Chang I C, Chen P C, Tsai M C, et al. Large-scale synthesis of uniform Cu_2O nanocubes with tunable sizes by in-situ nucleation [J]. CrystEngComm, 2013, 15: 2363-2366.

[37] Nikam A V, Arulkashmir V, Krishnamoorthy K, et al. pH-Dependent single-step rapid synthesis of CuO and Cu_2O nanoparticles from the same precursor [J]. Cryst Growth Des, 2014, 1: 4329-4334.

[38] Cao Y Y, Xu Y Y, Hao H Y, et al. Room temperature additive-free synthesis of uniform Cu_2O nanocubes with tunable size from 20nm to 500nm and photocatalytic property [J]. Mater Lett, 2014, 114: 88-91.

[39] Kumar S, Parlett C M A, Isaacs M A, et al. Facile synthesis of hierarchical Cu_2O nanocubes as visible light photocatalysts [J]. Appl Catal B, 2016, 189: 226-232.

[40] Karthikeyan S, Kumar S, Durndell L J, et al. Size-dependent visible light photocatalytic performance of Cu$_2$O nanocubes [J]. ChemCatChem, 2018, 10: 3554-3563.

[41] Xu H L, Wang W Z, Zhu W. Shape evolution and size-controllable synthesis of Cu$_2$O octahedra and their morphology-dependent photocatalytic properties [J]. J Phys Chem B, 2006, 110: 13829-13834.

[42] Siegfried M J, Choi K S. Directing the architecture of cuprous oxide crystals during electrochemical growth [J]. Angew Chem Int Ed, 2005, 44: 3282-3287.

[43] Kuo C H, Huang M H, Facile synthesis of Cu$_2$O nanocrystals with systematic shape evolution from cubic to octahedral structures [J]. J Phys Chem C, 2008, 112: 18355-18360.

[44] Pang H, Gao, Lu Q Y, Glycine-assisted double-solvothermal approach for various cuprous oxide structures with good catalytic activities [J]. Cryst Eng Comm, 201, 12: 406-412.

[45] Ho J Y, Huang M H. Synthesis of submicrometer-sized Cu$_2$O crystals with morphological evolution from cubic to hexapod structures and their comparative photocatalytic activity [J]. J Phys Chem C, 2009, 11: 14159-14164.

[46] Liang X, Gao L, Yang S, et al. Facile synthesis and shape evolution of single-crystal cuprous oxide [J]. Adv Mater, 2009, 21: 2068-2071.

[47] Yao K X, Yin X M, Wang T H, et al. Synthe, self-assembly, disassembly, and reassembly of two types of Cu$_2$O nanocrystals unifaceted with {001} or {110} planes [J]. J Am Chem Soc, 2010, 132: 6131-6144.

[48] Huang W C, Lyu L M, Yang Y C, et al. Synthesis of Cu$_2$O nanocrystals from cubic to rhombic dodecahedral structures and their comparative photocatalytic activity [J]. J Am Chem Soc, 2011, 134:1261-1267.

[49] Wang X P, Jiao S H, Wu D P, et al. A facial strategy for crystal engineering of Cu$_2$O polyhedrons with high-index facets [J]. Cryst Eng Comm, 2013, 15: 1849-1852.

[50] Chen K, Sun C, Song S, et al. Polymorphic crystallization of Cu$_2$O compound [J]. Cryst Eng Comm, 2014, 16: 5257-5267.

[51] Liu H, Miao W, Zhang Z, et al. Controlled synthesis of different shapes of Cu$_2$O via γ-irradiation [J]. Cryst Growth Des, 2009, 9: 1733-1740.

[52] Shang Y, Shao Y M, Zhang D F, et al. Recrystallization-induced self-assembly for the growth of Cu$_2$O superstructures [J]. Angew Chem Int Ed, 2014, 53: 11514-11518.

[53] Jiao S, Xu L, Jiang K, et al. Well-defined non-spherical copper sulfide mesocages with single-crystalline shells by shape-controlled Cu_2O crystal templating [J]. Adv Mater, 2006, 18: 1174-1177.

[54] Caruso F, Caruso R A, Mohwald H, Production of hollow microspheres from nanostructured composite particles [J]. Chem Mater, 1999, 11: 3309-3314.

[55] Wang L, Sasaki T, Ebina Y, et al. Fabrication of controllable ultrathin hollow shells by layer-by-layer assembly of exfoliated titania nanosheets on polymer templates[J]. Chem Mater, 2002: 14, 4827.

[56] Tsai Y H, Chiu C Y, Huang M H. Fabrication of diverse Cu_2O nanoframes through face-selective etching Fabrication of diverse Cu_2O nanoframes through face-selective etching [J]. J Phys Chem C, 2013, 117: 24611-24617.

[57] Xu Y Y, Jiao X L, Chen D R. PEG-assisted preparation of single-crystalline Cu_2O hollow nanocubes [J]. J Phys Chem C, 2008, 112: 16769-16773

[58] Zhong Z, Fang Y, Lu W, et al. Coherent single charge transport in molecular-scale silicon nanowires [J]. Nano Lett, 2005, 5: 1143-1146.

[59] Hu C C, Chang K H, Lin M C, et al. Design and tailoring of the nanotubular arrayed architecture of hydrous RuO_2 for next generation supercapacitors [J]. Nano Lett, 2006, 6: 2690-2695.

[60] Arnold M S, Avouris P, Pan Z W, et al. Field-effect transistors based on single semiconducting oxide nanobelts [J]. J Phys Chem B, 2003, 107: 659-663.

[61] Tan Y, Xue X, Peng X, et al. Controllable fabrication and electrical performance of single crystalline Cu_2O nanowires with high aspect ratios [J]. Nano Lett, 2007, 7: 3723-3728.

[62] Wang W Z, Wang G G, Wang X S, et al. Synthesis and characterization of Cu_2O nanowires by a novel reduction route [J]. Adv Mater, 2002: 14, 67-69.

[63] Ren Y, Ma Z, Bruce P G. Transformation of mesoporous Cu/Cu_2O into porous Cu_2O nanowires in ethanol [J]. Cryst Eng Comm, 2012, 14: 2617-2620.

[64] Cao M H, Hu C W, Wang Y H, et al. A controllable synthetic route to Cu, Cu_2O and CuO nanotubes and nanorods [J]. Chem Commun, 2003, 39: 1884-1885.

[65] Hacialioglu S, Men F, Jin S, Facile and mild solution synthesis of Cu_2O nanowires and nanotubes driven by screw dislocations [J]. Chem Commun, 2012, 48: 1174-1176.

[66] Guan L, Pang H, Wang J, et al. Fabrication of novel comb-like Cu_2O nanorod-based

structures through an interface etching method and their application as ethanol sensors [J]. Chem Commun, 2010, 46: 7022-7024.

[67] Musselman K P, Mulholland G J, Robinson A P, et al. MacManus-Driscoll, Low-temperature synthesis of large-area, free-standing nanorod arrays on ITO/glass and other conducting substrates [J]. Adv Mater, 2008, 20: 4470-4475.

[68] Ju H K, Lee J K, Lee J, et al. Fast and selective Cu_2O nanorod growth into anodic alumina templates via electrodeposition [J]. Curr Appl Phys, 2012, 12:60-63.

[69] Haynes K M, Perry C M, Rivas M, et al. Templated electrodeposition and photocatalytic of cuprous oxide nanorod arrays [J]. ACS Appl Mater Inter, 2015, 7: 830-837.

[70] Li H J, Zhou Y, Tu W, et al. Cu_2O/TiO_2 through a direct Z-scheme: protecting Cu_2O from photocorrosion [J]. Adv Funct Mater, 2015, 25: 998-1013.

[71] Liu L, Lin S, Hu J, et al. Plasmon-enhanced photocatalytic properties of nano Ag@AgBr on single-crystalline octahedral Cu_2O {111} microcrystals composite photocatalyst et al [J]. Appl Surf Sci, 2015, 330: 94-103.

[72] Aguirre M E, Zhou R, Eugene A J, et al. Cu_2O/TiO_2 heterostructures for CO_2 reduction through a direct Z-scheme: protecting Cu_2O from photocorrosion [J]. Appl Catal B, 2017, 217: 485-493.

[73] Ma J, Wang K, Li L, et al. Visible-light photocatalytic decolorization of orange II on Cu_2O/ZnO nanocomposites [J]. Ceram Int, 2015, 41:2050-2056.

[74] Kumar A, Sharma G, Al-Muhtaseb Aa H, et al. Quaternary magnetic $BiOCl/g-C_3N_4/Cu_2O/Fe_3O_4$ nano-junction for visible light and solar powered degradation of sulfamethoxazole from aqueous environment [J]. Chem Eng J, 2018, 334:462-478.

[75] Wang D, Saleh N B, Sun W, et al. Next-generation multifunctional carbon-metal nanohybrids for energy and environmental applications [J]. Environ Sci Technol, 2019, 53: 7265-7287.

[76] Wang P, Huang B B, Qin X Y, et al. Ag@ AgCl: a highly efficient and stable photocatalyst active under visible light [J]. Angew Chem Int Ed, 2008, 47: 7931.

[77] Hayes J R, Nyce G W, Kuntz J D, et al. Synthesis of bi-modal nanoporous Cu, CuO and Cu_2O monoliths with tailored porosity [J]. Nanotechnology, 2007, 18: 275602.

[78] Liu Y, Chu Y, Zhuo Y J, et al. Controlled synthesis of various hollow Cu nano/microstructures via a novel reduction route[J]. Adv Funct Mater, 2007, 17: 933-938.

[79] Lee Y H, Leu I C, Wu M T, et al. Fabrication of Cu/Cu$_2$O composite nanowire arrays on Si via AAO template-mediated electrodeposition [J]. J Alloys Compd, 2007, 427: 213-218.

[80] Teo J J, Chang Y, Zeng H C. Fabrications of hollow nanocubes of Cu$_2$O and Cu via reductive self-assembly of CuO nanocrystals [J]. Langmuir, 2006, 22: 7369-7377.

[81] Zheng H M, Liu X H, Yang S B, et al. New approach for preparation of ultrafine Cu particles and shell/core compounds of Cu/CuO and Cu/Cu$_2$O [J]. J Mater Sci, 2005, 40: 1039-1041.

[82] Sun S, Kong C, You H, et al. Facet-selective growth of Cu-Cu$_2$O heterogeneous architectures [J]. CrystEngComm, 2012, 14: 40-43.

[83] Chen W, Fan Z, Lai Z. Synthesis of core-shell heterostructured Cu/Cu$_2$O nanowires monitored by in situ XRD as efficient visible-light photocatalysts [J]. J Mater Chem A, 2013, 1: 13862-13868.

[84] Ai Z, Zhang L, Lee S, et al. Interfacial hydrothermal synthesis of Cu@ Cu$_2$O core-shell microspheres with enhanced visible-light-driven photocatalytic activity [J]. J Phys Chem C, 2009, 113: 20896-20902.

[85] Hou W, Cronin S B. A review of surface plasmon resonance enhanced photocatalysis [J]. Adv Funct Mater, 2013, 23: 1612-1619.

[86] Wang P, Huang B, Dai Y, et al. Plasmonic photocatalysts:harvesting visible light with noble metal nanoparticles [J]. Phys Chem Chem Phys, 2012, 14: 9813-9825.

[87] Cushing S K, Li J, Meng F, et al. Photocatalytic activity enhanced by plasmonic resonant energy transfer from metal to semiconductor[J]. J Am Chem Soc, 2012, 134: 15033-15041.

[88] Xiao F. Layer-by-Layer self-assembly construction of highly ordered metal-TiO$_2$ nanotube arrays heterostructures (M/TNTs, M = Au, Ag, Pt) with tunable catalytic activities [J]. J Phys Chem C, 2012, 116: 16487-16498.

[89] Zhang W, Yang X, Zhu Q, et al. One-pot room temperature synthesis of Cu$_2$O/Ag composite nanospheres with enhanced visible-light-driven photocatalytic performance [J]. Ind Eng Chem Res, 2014, 53:16316-16323.

[90] Lee C, Shin K, Lee Y J, et al. Effects of shell thickness on Ag-Cu$_2$O core-shell nanoparticles with bumpy structures for enhancing photocatalytic activity and stability [J]. Catal Today, 2018, 303: 313-319.

[91] Xiong J, Li Z, Chen J, et al. Facile synthesis of highly efficient one-dimensional plasmonic photocatalysts through Ag@Cu$_2$O core-shell heteronanowires [J]. ACS Appl Mater Inter, 2014, 6: 15716-15725.

[92] Hu Z, Mi Y, Ji Y, et al. Multiplasmon modes for enhancing the photocatalytic activity of Au/Ag/Cu$_2$O core-shell nanorods [J]. Nanoscale, 2019, 11: 16445-16454.

[93] Sharma K, Maiti K, Kim N H, et al. Green synthesis of glucose-reduced graphene oxide supported Ag-Cu$_2$O nanocomposites for the enhanced visible-light photocatalytic activity [J]. Compos Part B-Eng, 2018, 138: 35-44.

[94] Wei Q, Wang Y, Qin H, et al. Construction of rGO wrapping octahedral Ag-Cu$_2$O heterostructure for enhanced visible light photocatalytic activity [J]. Appl Catal B, 2018, 227:132-144.

[95] Liu X W. Selective growth of Au nanoparticles on {111} facets of Cu$_2$O microcrystals with an enhanced electrocatalytic property [J]. Langmuir, 2011, 27: 9100-9104.

[96] Zhu H, Du M, Yu D, et al. A new strategy for the surface-free-energy-distribution induced selective growth and controlled formation of Cu$_2$O-Au hierarchical heterostructures with a series of morphological evolutions[J]. J Mater Chem A, 2013, 1: 919-929.

[97] Mahmoud M A, Qian W M. Sayed Following charge separation on the nanoscale in Cu$_2$O-Au nanoframe hollow nanoparticles [J]. Nano Lett, 2011, 11: 3285-3289.

[98] Wang W C, Lyu L M. Huang M H. Investigation of the effects of polyhedral gold nanocrystal morphology and facets on the formation of Au-Cu$_2$O core-shell heterostructures [J]. Chem Mater, 2011, 23: 2677-2684.

[99] Miller E B, Zahran E M, Knecht M R, et al. Metal oxide semiconductor nanomaterial for reductive debromination: visible light degradation of polybrominated diphenyl ethers by Cu$_2$O@ Pd nanostructures [J]. Appl Catal B, 2017, 213: 147-154.

[100] Heng B, Xiao T, Tao W, et al. Zn doping induced shape evolution of microcrystals: the case of cuprous oxide [J]. Cryst Growth Des, 2012, 12: 3998-4005.

[101] Yu X, Zhang J, Zhang J, et al. Photocatalytic degradation of ciprofloxacin using Zn-doped Cu$_2$O particles: analysis of degradation pathways and intermediates [J]. Chem Eng J, 2019, 374: 316-327.

[102] Jiang T, Xie T, Chen L, et al. Carrier concentration dependent electron transfer in Cu$_2$O/ZnO nanorod arrays and their photocatalytic performance [J]. Nanoscale,

2013, 5: 2938-3944.

[103] Deo M, Shinde D, Yengantiwar A, et al. Cu_2O/ZnO hetero-nanobrush: hierarchical assembly, field emission and photocatalytic properties [J]. J Mater Chem, 2012, 22:17055-17062.

[104] Zou X, Fan H, Tian Y, et al. Synthesis of Cu_2O/ZnO hetero-nanorod arrays with enhanced visible light-driven photocatalytic activity [J]. CrystEngComm, 2014, 16: 1149-1156.

[105] Kandjani A E, Sabri Y M, Periasamy S R, et al. Controlling core/shell formation of nanocubic p-Cu_2O/n-ZnO toward enhanced photocatalytic performance [J]. Langmuir, 2015, 31: 10922-10930.

[106] Wu S C, Tan C S, Huang M H. Strong facet effects on interfacial charge transfer revealed through the examination of photocatalytic activities of various Cu_2O-ZnO heterostructures [J]. Adv Funct Mate, 2017, 27: 1604635.

[107] Ren S, Zhao G, Wang Y, et al. Enhanced photocatalytic performance of sandwiched ZnO@Ag@Cu_2O nanorod films: the distinct role of AgNPs in the visible light and UV region [J]. Nanotechnology, 2015, 26:125403.

[108] Liu H R, Yang J H, Zhang Y Y, et al. Prediction of $(TiO_2)_x(Cu_2O)_y$ alloys for efficient photoelectrochemical water splitting [J]. Phys Chem Chem Phys, 2013, 15: 1778-1781.

[109] Hou Y, Li X Y, Zhao Q D, et al. Fabrication of Cu_2O/TiO_2 nanotube heterojunction arrays and investigation of its photoelectrochemical behavior [J]. Appl Phys Lett, 2009, 95: 093108.

[110] Huang L, Peng F, Wang H, et al. Preparation and characterization of Cu_2O/TiO_2 nano-nano heterostructure photocatalysts [J]. Catal Commun, 2009, 10: 1839-1843.

[111] Wang M, Sun L, Lin Z, et al. p-n Heterojunction photoelectrodes composed of Cu_2O-loaded TiO_2 nanotube arrays with enhanced photoelectrochemical and photoelectrocatalytic activities [J]. Energ Environ Sci, 2013, 6: 1211-1220.

[112] Liu L, Yang W, Sun W, et al. Creation of Cu_2O@TiO_2 composite photocatalysts with p-n heterojunctions formed on exposed Cu_2O facets, their energy band alignment study, and their enhanced photocatalytic activity under illumination with visible light [J]. ACS appl Mater Inter, 2015, 7: 1465-1476.

[113] Liu L, Gu X, Sun C, et al. In situ loading of ultra-small Cu_2O particles on TiO_2

nanosheets to enhance the visible-light photoactivity [J]. Nanoscale, 2012, 4: 6351-6359.

[114] Yang L, Luo S, Li Y, et al. High efficient photocatalytic degradation of p-nitrophenol on a unique Cu_2O/TiO_2 p-n heterojunction network catalyst [J]. Environ Sci Technol, 2010, 44: 7641-7646.

[115] Fu J, Cao S, Yu J. Dual Z-scheme charge transfer in TiO_2-Ag-Cu_2O composite for enhanced photocatalytic hydrogen generation [J]. J Mater, 2015, 1: 124-133.

[116] Nakaoka K, Ueyama J, Ogura K. Photoelectrochemical behavior of electrodeposited CuO and Cu_2O thin films on conducting substrates [J]. J Electrochem Soc, 2004, 151: 661-665.

[117] Jiang D, Xue J, Wu L, et al. Photocatalytic performance enhancement of CuO/Cu_2O heterostructures for photodegradation of organic dyes: effects of CuO morphology [J]. Appl Catal B, 2017, 211:199-204.

[118] Yu H, Yu J, Liu S, et al. Template-free hydrothermal synthesis of CuO/Cu_2O composite hollow microspheres [J]. Chem Mater, 2007, 19: 4327-4334.

[119] Morales-Guio C, Liardet L, Mayer M T, et al. Photoelectrochemical hydrogen production in alkaline solutions using Cu_2O coated with earth-abundant hydrogen evolution catalysts [J]. Angew Chem Int Ed, 2015, 54: 664-667.

[120] Liu X, Chen J, Liu P, et al. Controlled growth of CuO/Cu_2O hollow microsphere composites as efficient visible-light-active photocatalysts [J]. Appl Catal A, 2016, 52:34-41.

[121] Chen J, Liu X, Zhang H, et al. Soft-template assisted synthesis of mesoporous CuO/Cu_2O composite hollow microspheres as efficient visible-light photocatalyst [J]. Mater Lett, 2016, 182: 47-51.

[122] Yurddaskal M, Dikici T, Celik E, Effect of annealing temperature on the surface properties and photocatalytic efficiencies of Cu_2O/CuO structures obtained by thermal oxidation of Cu layer on titanium substrates [J]. Ceram Int, 2016, 42, 17749-17753.

[123] Wang P, Wang J, Wang X, et al. Cu_2O-rGO-CuO composite: an effective Z-schemevisible-light photocatalyst [J]. Curr Nanosci, 2015, 11: 462-469.

[124] Ma H, Liu Y, Fu Y, et al. Improved photocatalytic activity of copper heterostructure composites (Cu-Cu_2O-CuO/AC) prepared by simple carbothermal reduction [J].

Aust J Chem, 2014, 67:749-756.

[125] Li H, Su Z, Hu S, et al. Free-standing and flexible Cu/Cu$_2$O/CuO heterojunction net: a novel material as cost-effective and easily recycled visible-light photocatalyst [J]. Appl Catal B, 2017, 207: 134-142.

[126] Ajmal A, Majeed I, Malik R N, et al. Photocatalytic degradation of textile dyes on Cu$_2$O-CuO/TiO$_2$ anatase powders [J]. J Environ Chem Eng, 2016, 4: 2138-2146.

[127] Zhao Y F, Yang Z Y, Zhang Y X, et al. Cu$_2$O decorated with cocatalyst MoS$_2$ for solar hydrogen production with enhanced efficiency under visible light [J]. J Phys Chem C, 2014, 118: 14238-14245.

[128] Lakhera S K, Watts A, Hafeez H Y, et al. Interparticle double charge transfer mechanism of heterojunction α-Fe$_2$O$_3$/Cu$_2$O mixed oxide catalysts and its visible light photocatalytic activity [J]. Catal Today, 2018, 300: 58-70.

[129] Tian Q, Wu W, Sun L, et al. Tube-like ternary α-Fe$_2$O$_3$@SnO$_2$@Cu$_2$O sandwich heterostructures: synthesis and enhanced photocatalytic properties [J]. ACS Appl Mater Inter, 2014, 6:13088-13097.

[130] Li F, Dong B. Construction of novel Z-scheme Cu$_2$O/graphene/α-Fe$_2$O$_3$ nanotube arrays composite for enhanced photocatalytic activity [J]. Ceram Int, 2017, 43: 16007-16012.

[131] Shen H, Liu G, Yan X, et al. All-solid-state Z-scheme system of RGO-Cu$_2$O/Fe$_2$O$_3$ for simultaneous hydrogen production and tetracycline degradation [J]. Mater Today Energy, 2017, 5: 312-319.

[132] Luo Y, Huang Q, Li B, et al. Synthesis and characterization of Cu$_2$O-modified Bi$_2$O$_3$ nanospheres with enhanced visible light photocatalytic activity [J]. Appl Surf Sci, 2015, 357: 1072-1079.

[133] Shen H, Wang J, Jiang J, et al. All-solid-state Z-scheme system of RGO-Cu$_2$O/Bi$_2$O$_3$ for tetracycline degradation under visible-light irradiation [J]. Chem Eng J, 2017, 313: 508-517.

[134] Wei S, Ma Y, Chen Y, et al. Fabrication of WO$_3$/Cu$_2$O composite films and their photocatalytic activity [J]. J Hazard Mater, 2011, 194: 243-249.

[135] Li Z P, Wen Y Q, Shang J P, et al. Magnetically recoverable Cu$_2$O/Fe$_3$O$_4$ composite photocatalysts: fabrication and photocatalytic activity [J]. Chin Chem Lett, 2014, 25:287-291.

[136] Bi Y, Ouyang S, Umezawa N, et al. Facet effect of single-crystalline Ag_3PO_4 sub-microcrystals on photocatalytic properties [J]. J Am Chem Soc, 2011, 133: 6490-6492.

[137] Li Z, Dai K, Zhang J, et al. Facile synthesis of novel octahedral Cu_2O/Ag_3PO_4 composite with enhanced visible light photocatalysis [J]. Mater Lett, 2017, 206: 48-51.

[138] Hou G, Zeng X, Gao S. Fabrication and photocatalytic activity of core@shell Ag_3PO_4@ Cu_2O heterojunction [J]. Mater Lett, 2019, 238: 116-120.

[139] Kakuta N, Goto N, Ohkita H, et al. Silver bromide as a photocatalyst for hydrogen generation from CH_3OH/H_2O solution [J]. J Phys Chem B, 1999, 103: 5917-5919.

[140] Hu X, Zhou X, Wang R, et al. Characterization and photostability of Cu_2O-Ag-$AgBr/Al_2O_3$ for the degradation of toxic pollutants with visible-light irradiation[J]. Appl Catal B, 2014, 154: 44-50.

[141] He J, Shao D W, Zheng L C, et al. Construction of Z-scheme Cu_2O/Cu/AgBr/Ag photocatalyst with enhanced photocatalytic activity and stability under visible light [J]. Appl Catal B, 2017, 203: 917-926.

[142] Xu Z K, Han L, Hu P, et al. Facile synthesis of small Ag@AgCl nanoparticles via a vapor diffusion strategy and their highly efficient visible-light driven photocatalytic performance [J]. Catal Sci Technol, 2014, 4: 3615-3619.

[143] Kim T G, Yeon D H, Kim T, et al. Silver silicates with three dimensional d^{10}-d^{10} interactions as visible light active photocatalysts for water oxidation[J]. Appl Phys Lett, 2013, 103: 043904.

[144] Long M C, Cai W M. Advanced nanoarchitectures of silver/silver compound composites for photochemical reactions [J]. Nanoscale, 2014, 6:7730-7742.

[145] An C H, Peng S N, Sun Y G. Facile synthesis of sunlight-driven AgCl:Ag plasmonic nanophotocatalyst [J]. Adv Mater, 2010, 22: 2570-2574.

[146] Li H Y, Sun Y J, Cai B, et al. Hierarchically Zscheme photocatalyst of Ag@AgCl decorated on $BiVO_4$ {040} with enhancing photoelectrochemical and photocatalytic performance [J]. Appl Catal B, 2015, 170: 206-214.

[147] Wang P, Yu C D, Ding J J, et al. Facile synthesis and improved photocatalytic performance of Ag-AgCl photocatalyst by loading basic zinc carbonate [J]. J Alloys Compd, 2018, 752: 238-246.

[148] Lou S, Wang W, Wang L, et al. In-situ oxidation synthesis of $Cu_2O/Ag/AgCl$ microcubes with enhanced visible-light photocatalytic activity[J] J Alloys Compd, 2019, 781: 508-514.

[149] He R, Cao S W, Zhou P, et al. Recent advances in visible light Bi-based photocatalysts [J]. Chin J Catal, 2014, 35: 989-1007.

[150] García-Pérez U M, Martínez-de la Cruz A, Sepúlveda-Guzmán S, et al. Low temperature synthesis of $BiVO_4$ powders by Pluronic-assisted hydrothermal method: effect of the surfactant and temperature on the morphology and structural control [J]. Ceram Int, 2014, 40:4631-4638.

[151] Li H, Hong W, Cui Y, et al. Enhancement of the visible light photocatalytic activity of $Cu_2O/BiVO_4$ catalysts synthesized by ultrasonic dispersion method at room temperature [J]. Mat Sci Eng B: Solid, 2014, 181: 1-8.

[152] Yuan Q, Chen L, Xiong M, et al. $Cu_2O/BiVO_4$ heterostructures: synthesis and application in simultaneous photocatalytic oxidation of organic dyes and reduction of Cr(Ⅳ) under visible light [J]. Chem Eng J, 2014, 255: 394-402.

[153] Wang W, Huang X, Wu S, et al. Preparation of p-n junction $Cu_2O/BiVO_4$ heterogeneous nanostructures with enhanced visible-light photocatalytic activity [J]. Appl Catal B, 2013, 134: 293-301.

[154] Deng Y, Tang L, Zeng G, et al. Plasmonic resonance excited dual Z-scheme $BiVO_4/Ag/Cu_2O$ nanocomposite: synthesis and mechanism for enhanced photocatalytic performance in recalcitrant antibiotic degradation[J]. Environ Sci Nano, 2017, 4: 1494-1511.

[155] Chen W, Liu T Y, Huang T, et al. In situ fabrication of novel Z-scheme Bi_2WO_6 quantum dots/$g-C_3N_4$ ultrathin nanosheets heterostructures with improved photocatalytic activity [J]. Appl Surf Sci, 2015, 355: 379-387.

[156] Girish Kumar S, Koteswara Rao K S R. Tungsten-based nanomaterials (WO_3, Bi_2WO_6): modifications related to charge carrier transfer mechanisms and photocatalytic applications [J]. Appl Surf Sci, 2015, 355: 939-958.

[157] Liu L, Ding L, Liu Y, et al. Enhanced visible light photocatalytic activity by Cu_2O-coupled flower-like Bi_2WO_6 structures [J]. Appl Surf Sci, 2016, 364: 505-515.

[158] Li J, Yuan H, Zhu Z. Fabrication of $Cu_2O/Au/BiPO_4$ Z-scheme photocatalyst to

improve the photocatalytic activity under solar light [J]. J Mol Catal A, 2015, 410: 133-139.

[159] Xia Y, He Z, Yang W, et al. Effective charge separation in BiOI/Cu$_2$O composites with enhanced photocatalytic activity [J]. Mater Res Express, 2018, 5:025504.

[160] Cui W, An W, Liu L, et al. Novel Cu$_2$O quantum dots coupled flower-like BiOBr for enhanced photocatalytic degradation of organic contaminant [J]. J Hazard Mater, 2014, 280: 417-427.

[161] Tian Y, Chang B, Fu J, et al. Graphitic carbon nitride/Cu$_2$O heterojunctions: Preparation, characterization, and enhanced photocatalytic activity under visible light [J]. J Solid State Chem, 2014, 212: 1-6.

[162] Li D, Zan J, Wu L, et al. Heterojunction tuning and catalytic efficiency of g-C$_3$N$_4$-Cu$_2$O with glutamate [J]. Ind Eng Chem Res, 2019, 58: 4000-4009.

[163] Liu L, Qi Y, Hu J, et al. Stable Cu$_2$O@ g-C$_3$N$_4$ core@ shell nanostructures: Efficient visible-light photocatalytic hydrogen evolution [J]. Mater Lett, 2015, 158: 278-281.

[164] Yan X, Xu R, Guo J J, et al. Enhanced photocatalytic activity of Cu$_2$O/g-C$_3$N$_4$ heterojunction coupled with reduced graphene oxide three-dimensional aerogel photocatalysis [J]. Mater Res Bull, 2017, 96: 18-27.

[165] Liang S, Zhou Y, Cai Z, et al. Preparation of porous g-C$_3$N$_4$/Ag/Cu$_2$O: a new composite with enhanced visible-light photocatalytic activity [J]. Appl Organomet Chem, 2016, 30: 932-938.

[166] Zuo S, Chen Y, Liu W, et al. A facile and novel construction of attapulgite/Cu$_2$O/Cu/g-C$_3$N$_4$ with enhanced photocatalytic activity for antibiotic degradation [J]. Ceram Int, 2017, 43: 3324-3329.

[167] Zhang S, Chen Q, Wang Y, et al. Synthesis and photoactivity of CdS photocatalysts modified by polypyrrole [J]. Int J Hydrogen Energy, 2012, 37: 13030-13036.

[168] Choi H, Kamat P V. CdS nanowire solar cells: dual role of squaraine dye as a sensitizer and a hole transporter [J]. J Phys Chem C, 2013, 4: 3983-3991.

[169] Rengaraj S, Venkataraj S, Jee S H, et al. Cauliflower-like CdS microspheres composed of nanocrystals and their physicochemical properties[J]. Langmuir, 2011, 27: 352-358.

[170] Zhai J, Wang L, Wang D, et al. Enhancement of gas sensing properties of CdS

nanowire/ZnO nanosphere composite materials at room temperature by visible-light activation [J]. ACS Appl Mater Interfaces, 2011, 3: 2253-2258.

[171] Liu Y, Dong H, Jia H, et al. CdS nanowires decorated with Cu_2O nanospheres: Synthesis, formation process and enhanced photoactivity and stability [J]. J Alloys Compd, 2015, 644: 159-164.

[172] Lin J, Lin J, Zhu Y F. Controlled synthesis of the $ZnWO_4$ nanostructure and effects on the photocatalytic performance [J]. Inorg Chem, 2007, 46: 8372-8378.

[173] Tian L, Rui Y, Sun K, et al. Surface decoration of $ZnWO_4$ nanorods with Cu_2O nanoparticles to build heterostructure with enhanced photocatalysis [J]. Nanomaterials, 2018, 8: 33.

[174] Yue Y, Zhang P, Wang W, et al. Enhanced dark adsorption and visible-light-driven photocatalytic properties of narrower-band-gap Cu_2S decorated Cu_2O nanocomposites for efficient removal of organic pollutants [J]. J Hazard Mater, 2020, 384: 121302.

[175] Balaji S, Selvan R K, Berchmans L J, et al. Enhanced dark adsorption and visible-light-driven photocatalytic properties of narrower-band-gap Cu_2S decorated Cu_2O nanocomposites for efficient removal of organic pollutants. Combustion synthesis and characterization of Sn^{4+} substituted nanocrystalline $NiFe_2O_4$ [J]. Mater Sci Eng B, 2005, 119: 119-124.

[176] Sohrabnezhad S, Rezaeimanesh M. Synthesis and characterization of novel magnetically separable $NiFe_2O_4$@ AlMCM-41-Cu_2O core-shell and its performance in removal of dye [J]. Adv Powder Technol, 2017, 28:3039-3048.

[177] He Z, Xia Y, Tang B, et al. Fabrication and photocatalytic property of magnetic $NiFe_2O_4$/Cu_2O composites [J]. Mater Res Express, 2017, 4: 095501.

[178] Yang S F, Niu C G, D. Huang W, et al. $SrTiO_3$ nanocubes decorated with Ag/AgCl nanoparticles as photocatalysts with enhanced visible-light photocatalytic activity towards the degradation of dyes, phenol and bisphenol A [J]. Environ Res: Nano, 2017, 4: 585-595.

[179] Guo J, Ouyang S, Li X P, et al. A new heterojunction Ag_3PO_4/Cr-$SrTiO_3$ photocatalyst towards efficient elimination of gaseous organic pollutants under visible light irradiation [J]. Appl Catal B, 2013, 134-135: 286-292.

[180] Xia Y M, Yu X Q, He Z M, et al. Synthesis of novel copper-based oxide nan-

ostructured film on copper via solution-immersion [J]. Ceram Int, 2017, 43: 14499-14503

[181] Xia Y, He Z, Hu K, et al. Fabrication of n-SrTiO$_3$/p-Cu$_2$O heterojunction composites with enhanced photocatalytic performance [J]. J Alloys Compds, 2018, 753:356-363.

[182] Zhao Q, Wang J L, Li Z P, et al. Heterostructured graphitic-carbon-nitride-nanosheets/copper(I) oxide composite as an enhanced visible light photocatalyst for decomposition of tetracycline antibiotics [J]. Sep Purif Technol, 2020, 250（11）: 117238-117245.

[183] Zhao Q, Wang J L, Li Z P, et al. Preparation of Cu$_2$O/exfoliated graphite composites with high visible light photocatalytic performance and stability [J]. Ceram Int, 2016, 42: 13273-13277.

[184] Zhao Q, Wang J L, Li Z P, et al. Two-dimensional Ti$_3$C$_2$T$_X$-nanosheets/Cu$_2$O composite as a high-performance photocatalyst for decomposition of tetracycline [J]. Carbon Resources Conversion, 2021, 4: 197-204.

[185] Zhao Q, Wang J L, Li Z P, et al, UiO-66-NH$_2$/Cu$_2$O composite as an enhanced visible light photocatalyst fordeco mposition of organic pollutants [J]. J Photochem Photobiol A: Chem, 2020, 399(8): 112625-112632.

[186] Zhao Q, Wang K W, Wang J L, et al, Cu$_2$O Nanoparticle hyper-cross-linked polymer composites for the visible-light photocatalytic degradation of methyl orange [J]. ACS Appl Nano Mater, 2019, 2(5): 2706-2712.

第 2 章

催化剂的制备与表征方法

- 2.1 试剂与仪器
- 2.2 Cu_2O 的制备
- 2.3 Cu_2O 基光催化剂的制备
- 2.4 光催化剂的表征方法

2.1 试剂与仪器

（1）实验试剂（见表 2.1）

表 2.1　实验试剂

名称	规格	生产单位
硫酸铜	AR	天津市化学试剂批发公司
氯化铜	AR	天津市北方化玻采购销售中心
氢氧化钠	AR	天津市风船化学试剂科技有限公司
无水乙醇	AR	天津市风船化学试剂科技有限公司
硝酸钠	AR	天津市大陆化学试剂厂
氯化铁	AR	天津市北方化玻采购销售中心
二甲醇缩甲醛	AR	天津市化学试剂批发公司
苯	AR	天津市化学试剂批发公司
二氯乙烷	AR	天津市化学试剂批发公司
甲醇	AR	天津市化学试剂批发公司
抗坏血酸	AR	上海市阿拉丁生化科技有限股份公司
盐酸羟胺	AR	天津市化学试剂批发公司
双氧水	AR	天津市红岩化学试剂厂
浓硫酸	AR	天津市红岩化学试剂厂
天然石墨	—	山东青岛莱西化工
N-CMK-3	—	南京先丰纳米材料科技有限公司
g-C_3N_4	—	南京先丰纳米材料科技有限公司
四环素	AR	上海源叶生物有限公司
甲基橙	AR	天津市化学试剂一厂

（2）实验仪器（见表 2.2）

表 2.2　实验仪器

名称	型号	生产单位
集热式恒温加热磁力搅拌器	DF-101S	巩义市予华仪器有限责任公司
电热恒温鼓风干燥箱	DHG-9070A	上海齐欣科学仪器有限公司

续表

名称	型号	生产单位
高速离心机	HC-2518	安徽中科中佳科学仪器有限公司
电子天平	FA2004	上海舜宇恒平科学仪器有限公司
数控超声波清洗器	SB-5200	宁波新芝生物科技股份有限公司
分光光度计	722s 型	上海棱光技术有限公司
氙灯光源系统（500W）	CEL-S500	北京中教金源科技有限公司
短弧氙灯系统（500W）	—	北京畅拓科技有限公司

2.2 Cu_2O 的制备

2.2.1 Cu_2O（菱方十二面体）的制备

取 20mL 无水乙醇加入到一个干燥的 250mL 烧杯中，随后再向其中加入 5mL $CuCl_2$ 溶液（0.5mol/L）和 83.4mL H_2O，然后用塑料薄膜进行密封，按一定的速度在恒温水浴锅中搅拌 30min，温度为 40℃；量取 30mL 无水乙醇和 9mL 1.0mol/L 的 NaOH 溶液，用恒压滴液漏斗滴加，控制不同的滴加速度，其中 CH_3CH_2OH 为每滴 3~4s、NaOH 溶液为每滴 6~7s；滴完后再加入 9.8mL 的 $NH_2OH·HCl$（0.5mol/L）溶液，继续搅拌 10min。然后静置 3h 后，除去上层清液，剩余溶液高速离心 10min，离心后沉淀部分用体积比为 1∶1 的 CH_3CH_2OH 和 H_2O 混合溶液洗涤 2~3 次，最后将沉淀放入表面皿中，在无水乙醇溶液中静置，放入 35℃烘箱中抽真空干燥，得到 Cu_2O 催化剂。

2.2.2 Cu_2O（立方体）的制备

在 500mL 烧杯中加入 1.5mL $CuSO_4$ 溶液（0.1mol/L）、135.75mL 蒸馏水，加入磁子并将烧杯放入 35℃恒温水浴中，用保鲜膜密闭烧杯口，搅拌 30min 后量取 5.25mL 氢氧化钠（1mol/L）向烧杯中滴加，烧杯中溶液变成浅蓝色，表明形成了 Cu(OH)$_2$ 沉淀。氢氧化钠滴加完后,量取 7.5mL 抗坏血酸(0.2mol/L)并以每滴 3~5s 的滴速将抗坏血酸加入烧杯,溶液呈现出亮黄色,将抗坏血酸全部加入烧杯后并用保鲜膜密封,继续搅拌 10min,然后静置 3h,之后除去上层清液,剩余溶液高速离心 10min,离心后沉淀用体积比为 1∶1 的 CH$_3$CH$_2$OH 和 H$_2$O 混合溶液洗涤 2~3 次,最后将沉淀放入表面皿中,在无水乙醇溶液中静置,放入 35℃烘箱中抽真空干燥得 Cu$_2$O 催化剂。

2.3 Cu$_2$O 基光催化剂的制备

2.3.1 Cu$_2$O/EG 光催化剂的制备

(1)膨胀石墨(EG)的制备

量取 40mL 98%硫酸,称取天然石墨 2g 加入到浓硫酸中,然后将其置于冰盐浴中进行冷却处理,待温度降至 0℃后,称取 2g 硝酸钠,搅拌同时均匀缓慢地加入到冷却溶液中,待反应 10min 后再慢慢加入 6g 高锰酸钾,在冰盐浴中搅拌反应 1h 后,撤去冰盐浴,以每秒 1 滴的速度滴加 125mL 蒸馏水和 100mL 热水,滴加 20mL 双氧水,溶液呈金黄色。冲洗滤液使其呈中性,抽滤,将滤饼于 65℃下烘干 4h,即制得氧化石墨。之后,在马弗炉中将氧化石墨于 1000℃下膨化,则制得所需膨胀石墨。

(2)Cu$_2$O/EG 的制备

干燥的烧杯(250mL)中,加入 20mL 无水乙醇、5mL CuCl$_2$ 溶液(0.5mol/L)、83.4mL H$_2$O 及 EG(0.1g),然后用塑料薄膜密封超声 30min,在 40℃水浴中搅拌 30min,量取 30mL 无水乙醇和 9mL 1.0mol/L 的 NaOH 溶液加入到恒压滴液漏斗中,控制不同的滴加速度滴加,其中 CH$_3$CH$_2$OH 为每滴 3~4s,NaOH 溶液为每滴 6~7s;滴完后再加入 9.8mL 的 NH$_2$OH·HCl(0.5mol/L)溶液,继续搅拌 10min。然后静置 3h 后,除去上层清液,剩余溶液高速离心 10min,离心后沉淀用体积比为 1∶1 的 CH$_3$CH$_2$OH 和 H$_2$O 混合溶液洗涤 2~3 次,再用无水乙醇洗涤 1 次,最后将沉淀物放入 35℃烘箱中抽真空干燥,得到 1% Cu$_2$O/EG 的催化剂。

其他比例的 Cu_2O/EG 催化剂制备步骤同上。

2.3.2　KAPs-B/Cu_2O 光催化剂的制备

（1）KAPs-B 的制备

采用外交联剂二甲醇缩甲醛（FDA）与各种芳香族单体通过傅-克反应交联的方法来制备超交联聚合物（HCPs）材料。在烧瓶中将单体（苯，1.56g），交联剂（FDA，4.56g）和催化剂（$FeCl_3$，9.75g）溶解于 1，2-二氯乙烷（DCE，20mL），在 45℃下加热 5h 形成网络，然后在 80℃加热 19h 完成缩合反应，生成微孔聚合物。固体产物用甲醇多次洗涤，直至滤液为无色，产品在甲醇里面采用索氏提取法提取 24h，然后在 60℃下真空干燥 24h 得到超交联苯基聚合物（KAPs-B）。

（2）KAPs-B/Cu_2O 催化剂的制备

如图 2.1 所示在干燥的烧杯（250mL）中，加入 20mL 无水乙醇、5mL $CuCl_2$ 溶液（0.5mol/L）、83.4mL H_2O 及 KAPs-B（0.007g），然后用塑料薄膜密封超声 30min，在 40℃水浴中搅拌 30min，量取 30mL 无水乙醇和 9mL 1.0mol/L 的 NaOH 溶液加入到恒压滴液漏斗中，控制不同的滴加速度滴加，

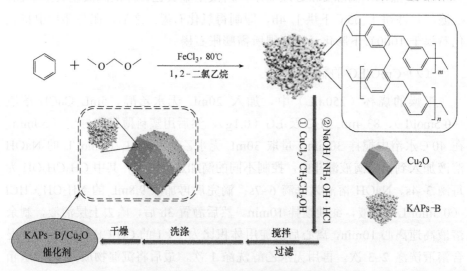

图 2.1　KAPs-B/Cu_2O 催化剂的制备

其中 CH_3CH_2OH 为每滴 3~4s、NaOH 溶液为每滴 6~7s；滴完后再加入 9.8mL 的 $NH_2OH \cdot HCl$（0.5mol/L）溶液，继续搅拌 10min。然后静置 3h 后，除去上层清液，剩余溶液高速离心 10min，离心后沉淀用体积比为 1：1 的 CH_3CH_2OH 和 H_2O 混合溶液洗涤 2~3 次，再用无水乙醇洗涤 1 次，最后将沉淀物放入 35℃烘箱中抽真空干燥，得到 7% KAPs-B/Cu_2O（质量分数）的催化剂。

其他比例 KAPs-B/Cu_2O 催化剂制备步骤同上。

2.3.3　UiO-66-NH_2/Cu_2O 光催化剂的制备

（1）UiO-66-NH_2 的制备

将 $ZrCl_4$（0.233g）和 2-氨基-1,4-苯二羧基酸（0.1812g）溶解于 50mL DMF 中，然后溶液被转移到 100mL 反应釜中。在 140℃下反应 24h。冷却至室温，产品经离心，乙醇洗涤三次去除 DMF，在 70℃下真空干燥 12h 得到黄色的 UiO-66-NH_2。

（2）UiO-66-NH_2/Cu_2O 光催化剂的制备

本文用浸渍法得到了复合材料。在一个 60mL 试剂瓶中加入 0.08g Cu_2O、20mL 乙醇和 0.02g UiO-66-NH_2，超声处理 60min 后，将混合物在 35℃下真空干燥 12h，得到 20%（质量分数）的 UiO-66-NH_2/Cu_2O 复合材料。同样，其他比例 UiO-66-NH_2/Cu_2O 催化剂制备步骤同上。

2.3.4　g-C_3N_4 纳米片/Cu_2O 光催化剂的制备

（1）g-C_3N_4 纳米片的制备

如图 2.2 所示，在 250mL 烧杯中，将 g-C_3N_4（1g）加入到浓硫酸（20mL，98%）中，然后将其置于冰盐浴中降温至 0℃后，将 1g 硝酸钠边搅拌边缓慢地加入到冷却溶液中，待反应 10min 后再慢慢加入 3g 高锰酸钾，在冰盐浴中继续搅拌 1h 后，撤去冰盐浴，以 1 滴/s 的速度滴加 125mL 蒸馏水，然后再以 1 滴/s 的速度滴加 20mL 双氧水。所得溶液用蒸馏水洗涤至中性，然后高速离心，在 35℃下真空干燥 12h 得到白色的 g-C₃N₄ 纳米片。

图 2.2　g-C₃N₄ 纳米片的制备

（2）g-C₃N₄ 纳米片/Cu₂O 的制备

在 500mL 烧杯中加入 1.5mL CuSO₄ 溶液（0.1mol/L）、135.75mL 蒸馏水和 g-C₃N₄ 纳米片（0.003g），超声处理 30min 后，在 35℃恒温水浴中，用保鲜膜密闭，搅拌 30min 后滴加 5.25mL 氢氧化钠（1mol/L），滴速为每滴 6~7s。然后量取 7.5mL 抗坏血酸（0.2mol/L）并以每滴 3~5s 的滴速将抗坏血酸加入烧杯，继续搅拌 10min 然后静置 3h 后。除去上层清液，剩余溶液高速离心 10min，离心后沉淀用体积比为 1∶1 的 CH₃CH₂OH-H₂O 混合溶液洗涤 3 次，再用无水乙醇洗涤 1 次，最后将沉淀物放入 35℃烘箱中抽真空干燥，得到质量分数为 30%的 g-C₃N₄ 纳米片/Cu₂O 催化剂。

其他比例 g-C₃N₄ 纳米片/Cu₂O 催化剂制备步骤同上。

2.3.5　Ti₃C₂Tx/Cu₂O 光催化剂的制备

（1）Ti₃C₂Tx 少层纳米片的制备

如图 2.3 所示，在 100mL 烧杯中，将 Ti₃C₂Tx（500mg）加入到浓硫酸（5mL，98%）中，然后将其置于冰盐浴中降温至 0℃后，将 0.5g 硝酸钠在搅拌条件缓慢地加入到冷却溶液中，待反应 10min 后再慢慢加入 1.5g 高锰酸钾，在冰盐浴中继续搅拌 1h 后，撤去冰盐浴，以 2~3 滴/s 的速度滴加 100mL 蒸馏水，所得溶液用蒸馏水洗涤至中性，然后高速离心，在 35℃下真空干燥 12h，得到白色的 Ti₃C₂T 少层纳米片。

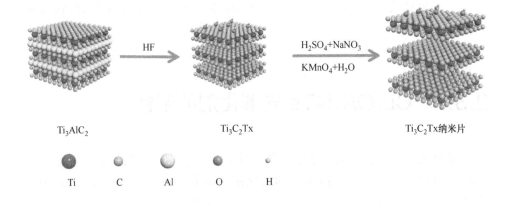

图 2.3　Ti_3C_2Tx 纳米片的制备

（2）2D Ti_3C_2Tx/Cu_2O 光催化剂的制备

在 250mL 烧杯中加入 0.75mL $CuSO_4$ 溶液（0.1mol/L）、67.8mL 蒸馏水和 g-C_3N_4 纳米片（3.5mg），超声处理 30min 后，在 35℃恒温水浴中，用保鲜膜密闭，搅拌 30min 后滴加 2.63mL 氢氧化钠（1mol/L），滴速为每滴 6~7s。然后量取 3.75mL 抗坏血酸（0.2mol/L）并以每滴 3~5s 的滴速将抗坏血酸加入烧杯，继续搅拌 10min 然后静置 3h，随后除去上层清液，剩余溶液高速离心 10min，离心后沉淀用体积比为 1∶1 的 CH_3CH_2OH-H_2O 混合溶液洗涤 3 次，再用无水乙醇洗涤 1 次，最后将沉淀物放入 35℃烘箱中抽真空干燥得 7% Ti_3C_2Tx/Cu_2O 的催化剂，其他比例 Ti_3C_2Tx/Cu_2O 光催化剂制备步骤同上。

2.3.6　N-CMK-3/Cu_2O 光催化剂的制备

分别向烧杯中加入 20mL 无水乙醇，随后再向其中加入 5mL $CuCl_2$ 溶液（0.5mol/L）和 83.4mL H_2O，再分别加入 0.02g 氮掺杂的有序介孔碳（N-CMK-3），然后用塑料薄膜进行密封，超声 30min，结束后取 30mL 无水乙醇和 9mL 1.0mol/L 的 NaOH 溶液分别滴加，其中无水乙醇、NaOH 溶液滴加速度分别为每滴 3~4s、每滴 6~7s；滴完后再加入 NH_2OH·HCl 溶液（9.8mL，0.5mol/L），持续搅拌 10min。静置 3h 后，高速离心，用无水乙醇和蒸馏水溶液（体积比为 1∶1）洗涤 3 次，放入 35℃烘箱中真空干燥,得到 1% Cu_2O/N-CMK-3 催化剂。

其他比例 N-CMK-3/Cu_2O 催化剂制备步骤同上。

2.3.7　Cu_2O/CNTs 光催化剂的制备

量取 83.4mL 的去离子水加入烧杯中，放置在 40℃的恒温水浴锅中，待温度恒定后，将 5mL 0.5mol/L 的氯化铜溶液和 10mL 的乙醇及 0.1g 的碳纳米管（CNTs）加入烧杯中，搅拌均匀后，加入 9mL 1.0mol/L 的氢氧化钠溶液，在加入的同时滴几滴乙醇，溶液变为浅蓝色后，在 5s 内加入 9.8mL 0.5mol/L 的盐酸羟胺，然后搅拌约 3min 后，在 40℃的水浴锅中静置 1.5h，之后搅拌约 15min，在室温下静置冷却后倒出上面溶液，用无水乙醇冲洗一次，放入 35℃烘箱中真空干燥得 1% Cu_2O/CNTs 催化剂。其他比例 Cu_2O/CNTs 催化剂制备步骤同上。

2.4　光催化剂的表征方法

2.4.1　结构表征

用 Bruker D8 衍射仪进行 X 射线粉末衍射（XRD）表征，Cu Kα 为辐射源，λ = 1.5418Å，扫描范围为 10°~80°，工作电压为 40kV。通过透射电镜（JEOLJEM, 200kV）和扫描电镜（TESCAN, 10kV）观察样品的形貌。用傅里叶变换红外光谱仪（CDS-2000 Bio-Rad FTS-165），KBr 压片来测定样品的红外谱图。采用 Thermo Fisher Scientific Escalab 250Xi 系统对 X 射线光电子能谱（XPS）进行表征。样品的比表面积由美国麦克仪器公司（ASAP-2020 PLUS HD88）物理吸附仪测定，通过吸附等温方程获得。在 Jena 分光光度计（SPECORD®210 PLUS）上获得样品的紫外-可见漫反射吸收光谱(UV-Vis DRS)，测试是以 $BaSO_4$ 作为参照。在含 KCl（0.1mol/L）和 $[Fe(CN)_6]^{3-/4-}$（2.5mmol/L）的溶液中进行了电化学阻抗谱（EIS）测试。在 Na_2SO_4（0.2mol/L）

溶液中进行光电流响应，以 Pt 为对电极，Ag/AgCl 为参比电极。

2.4.2 光催化性能评价

在模拟可见光照射下（氙灯 500W），取一个夹层石英容器，向其中加入 90mL（30mg/L）的四环素溶液和 100mg 的光催化剂，进行不断搅拌并发生暗反应 30min。取 4mL 四环素溶液放入比色皿中，用分光光度计测其吸光度 A_0，选择 λ_{max}=373nm；暗反应结束后，取 4mL 试样高速离心 5min，离心后取上层清液，测其吸光度。将溶液放在光照条件下进行反应，每隔 10min 进行一次取样、离心、测值。

根据下式计算四环素溶液的降解率（D）：

$$D = \frac{A_0 - A_t}{A_0} \times 100\%$$

式中，A_0、A_t 分别为光照前和光照 t 时间后的四环素溶液在 373nm 波长处的吸光度值，甲基橙溶液测定步骤同上，甲基橙的 λ_{max}=464nm。

2.4.3 活性组分

考察了不同捕获剂对 Cu_2O 基光催化剂性能的影响，通过引入异丙醇（IPA）、草酸铵（AO）和氮气氛围，对应捕获反应过程中产生的羟基自由基（HO·）、空穴（h^+）和超氧自由基（$O_2^{-·}$），以研究光催化降解甲基橙（MO）和四环素（TC）的机理。

第 3 章

p-n 型异质结光催化剂 g-C₃N₄/Cu₂O

3.1 引言
3.2 g-C₃N₄/Cu₂O 催化剂的结构与性能表征
3.3 光催化降解四环素性能测试
3.4 光催化活性组分测试及机理研究

3.1 引言

p-n 型异质结是一种有效的电荷收集或分离结构。通常,当 p 型半导体与 n 型半导体结合时,由于电荷-空穴(e^--h^+)的运动,会在界面处形成 p-n 型异质结,产生内部电场,e^--h^+ 会很快分离,e^- 转移到 n 型半导体的 CB,h^+ 转移到 p 型半导体的价带。p-n 型异质结具有以下优点:更有效的电荷分离、电荷快速转移到催化剂表面和更长的电荷载流子寿命。Zou 等人[1]研究表明 Cu_2O 与 ZnO 结合后,Cu_2O 的带隙变窄,太阳光谱的吸收范围可以扩大。此外,p-n 型异质结的形成有助于在可见光照射下分离光致电荷-空穴对,避免电荷复合。如图 3.1(a)所示,Cu_2O 导带(CB)中的光生电子可以迁移到 ZnO 纳米棒的 CB,而光生空穴从 ZnO 纳米棒的价带(VB)向相反方向迁移到 Cu_2O。Xia 等人[2]指出 n 型 $SrTiO_3$ 粒子与 p 型 Cu_2O 立方体结合后 [图 3.1(b)],根据 p-n 型异质结的形成理论,电荷载流子将在 Cu_2O 和 $SrTiO_3$ 之间扩散和转移。最后,在两个半导体的界面处形成 p-n 型异质结。在模拟的阳光下会产生大量的光生电荷载流子。Cu_2O 中形成的光生电子通过界面转移到 $SrTiO_3$ 上,光生空穴从 $SrTiO_3$ 转移到 Cu_2O 上,进一步增加了光生电子-空穴对的分离,从而增加了光生载流子的寿命。Cu_2O/TiO_2[3]、Cu_2O/Fe_2O_3[4]、

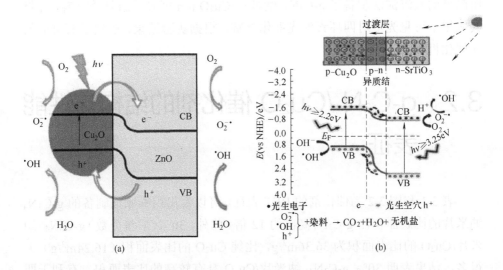

图 3.1 Cu_2O/ZnO(a)[1]和 $SrTiO_3/Cu_2O$ (b)[2]的 p-n 型异质结机理

WO_3/Cu_2O[5]、$Cu_2O/AgCl$[6]、$Cu_2O/AgBr$[7]、$Cu_2O/g-C_3N_4$[8]和CdS/Cu_2O[9]等复合物也可以用上述机理来解释。

石墨碳氮化物 $g-C_3N_4$ 是具有优良光催化性能的无金属高分子材料[10,11]。纳米片 $g-C_3N_4$ 的比表面积相对较大，可以为有机污染物的分解提供更丰富的反应位点，并能抑制光生 e^--h^+ 对的复合[12-19]。近年来，许多制备 $g-C_3N_4$ 纳米片的方法已经被广泛研究。研究者发现 $g-C_3N_4$ 与其他半导体形成异质结构是提高其光催化效率的有效途径[20-25]。例如，制作的 $g-C_3N_4/Cu_2O$ 异质结具有高化学稳定性、低毒性和高氧化能力。Tian 等人[8]通过水热法制备了 p-n 型异质结光催化剂 $g-C_3N_4/Cu_2O$，并将其应用于甲基橙的降解，其光催化活性明显高于纯 $g-C_3N_4$ 或 Cu_2O。Li 等人[26]通过水热合成法制备了 $g-C_3N_4/Cu_2O$，催化剂在制备过程中通过谷氨酸对催化剂进行修饰后，甲基橙的降解率从 80% 提高到 98%。因此，研究者认为利用谷氨酸可以增强异质结效应，提高光催化性能。Liu 等人[27]用溶剂热法和化学吸附法相结合的方法成功地在 Cu_2O 纳米球表面包覆了 $g-C_3N_4$。获得的 $Cu_2O@g-C_3N_4$ 核壳型复合材料对光催化产氢的性能明显增强。然而，在实际应用中，$g-C_3N_4/Cu_2O$ 复合材料的光催化效率还不够，需要进一步改进。特别是 $g-C_3N_4$ 的比表面积应进一步增加，以提供更丰富的活性位点，降低光生电子-空穴对的复合[28]。

在本研究中，采用膨胀法制备了比表面积大的 $g-C_3N_4$ 纳米薄片，然后用简单的共沉淀法制备了 $g-C_3N_4$ 纳米片/Cu_2O p-n 型异质结复合材料，并将其应用于可见光下对四环素的光催化降解。根据实验结果，提出了其可能的光催化机理[29]。

3.2　$g-C_3N_4/Cu_2O$ 催化剂的结构与性能表征

表 3.1 和图 3.2 为催化剂的 BET 表征。可以看出用膨胀法制备的 $g-C_3N_4$ 纳米片的比表面积比 $g-C_3N_4$ 高出约 12 倍。此外，30%（质量分数）$g-C_3N_4$ 纳米片/Cu_2O 的比表面积为 $36.36m^2/g$，比纯 Cu_2O 的比表面积（$16.24m^2/g$）大得多，结果表明 30% $g-C_3N_4$ 纳米片/Cu_2O 具有较高的比表面积，有利于四环素的吸附，从而提高光催化降解效率。

表 3.1　样品的比表面积

样本	比表面积/（m²/g）
g-C₃N₄	12.26
g-C₃N₄ 纳米片	150.16
Cu₂O	16.24
30% g-C₃N₄ 纳米片/Cu₂O	36.36

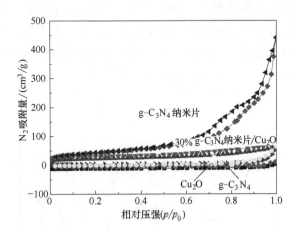

图 3.2　Cu$_2$O、g-C$_3$N$_4$、g-C$_3$N$_4$ 纳米片和 30%g-C$_3$N$_4$纳米片/Cu$_2$O 的 N$_2$ 吸附-解吸等温线

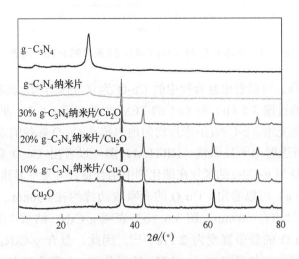

图 3.3　Cu$_2$O、g-C$_3$N$_4$、g-C$_3$N$_4$ 纳米片和 g-C$_3$N$_4$纳米片/Cu$_2$O 的 XRD 谱图

如图 3.3 所示，2θ 为 13.0°和 27.4°分别对应 g-C_3N_4 的 {100}和{002}晶面，该晶面与层间堆积结构有关[21]，如图所示，膨胀剥离后对应的{002}衍射强度明显降低，由此可知采用膨胀法剥离的 g-C_3N_4 纳米片为少层结构。2θ 为 29.3°、36.2°、42.1°、61.3°和 73.4°，对应 Cu_2O 的{110}、{111}、{200}、{220}和{311}晶面。此外，2θ 为 27.4°对应的{002}衍射峰也存在于 g-C_3N_4 纳米片/Cu_2O 复合物中，说明 g-C_3N_4 纳米片和 Cu_2O 成功复合。

图 3.4 为 g-C_3N_4 催化剂的 FTIR 光谱，其峰值在 1630cm^{-1} 附近被指定为 C=N 拉伸振动，峰值在 1243cm^{-1} 和 1320cm^{-1} 对应为芳香 C—N 的拉伸振动，吸收带在 805cm^{-1} 处属于三嗪环结构，对应于缩合的 C—N 杂环，宽带范围在 3000~3500cm^{-1} 对应未缩合的氨基和羟基基团[30]。

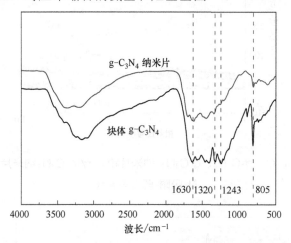

图 3.4 块体 g-C_3N_4 与 g-C_3N_4 纳米片的 FTIR 光谱

从图 3.5（a）可以看出复合物中的 Cu_2O 为立方体结构，其粒径在 100nm 左右。此外，通过图 3.5（b）和（c）的 TEM 图观察到 g-C_3N_4 纳米片呈典型的片层结构，表面光滑。g-C_3N_4 纳米片均匀地包覆在 Cu_2O 颗粒的表面，从图 3.5（d）的高分辨透射电子显微镜（HRTEM）图可以看出 Cu_2O 的晶面间距为 0.246nm，Cu_2O 与 g-C_3N_4 纳米片在催化剂中结合良好并形成异质结结构。

从图 3.6（a）可以看出，Cu_2O 的光吸收边缘约在 630nm，g-C_3N_4 纳米片的光吸收边缘约在 500nm。图 3.6（b）表明 g-C_3N_4 纳米片的禁带宽度为 2.68eV[31]，Cu_2O 的禁带宽度为 2.17eV[32]。因此，复合 g-C_3N_4 纳米片后可以提高 Cu_2O 催化剂的吸光度，g-C_3N_4 纳米片/Cu_2O 复合物对可见光的吸收强度增强，因此该催化剂可以获得更多的可见光，从而提高光催化性能。

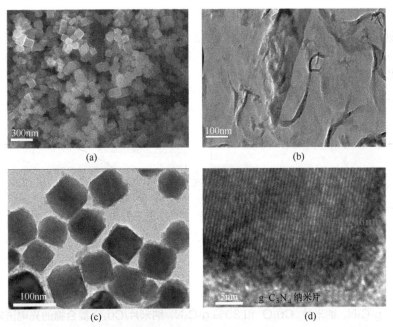

图 3.5 Cu$_2$O 的 SEM（a）、g-C$_3$N$_4$纳米片的 TEM（b）、30% g-C$_3$N$_4$纳米片/Cu$_2$O 复合物（质量分数）的 TEM（c）和 HRTEM 图像（d）

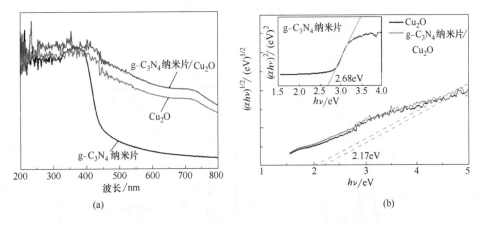

图 3.6 Cu$_2$O、g-C$_3$N$_4$ 纳米片和 g-C$_3$N$_4$ 纳米片/Cu$_2$O 的 UV-Vis 漫反射（DRS）光谱（a）与能带谱图（b）

（b）中插图为 g-C$_3$N$_4$ 纳米片能带谱

采用光电流响应和 EIS 图谱来评价电子-空穴对在光催化剂中的分离和转移能力。从图 3.7（a）可以看出，30% g-C$_3$N$_4$ 纳米片/Cu$_2$O 复合物的光电流值明显增强，表明光诱导的电子-空穴对在 g-C$_3$N$_4$ 纳米片上能更有效地分离。图 3.7（b）为催化剂的 EIS 图。从图可以看到，g-C_3N_4 纳米片/Cu_2O 复合物的圆半径比 Cu_2O 的小，表明复合材料具有比 Cu_2O 更快的界面电荷-电子转移速率[33]。因此，与纯 Cu_2O 相比，g-C_3N_4 纳米片/Cu_2O 复合材料具有更强的光诱导和传递载流子能力。

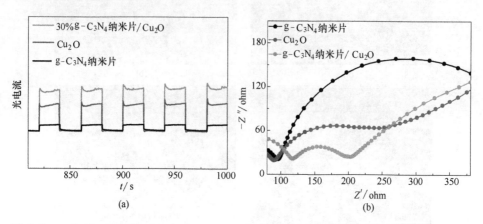

图 3.7　g-C_3N_4 纳米片、Cu_2O 和 30% g-C_3N_4 纳米片/Cu_2O 复合物的光电流响应信号（a）和 EIS 图谱（b）

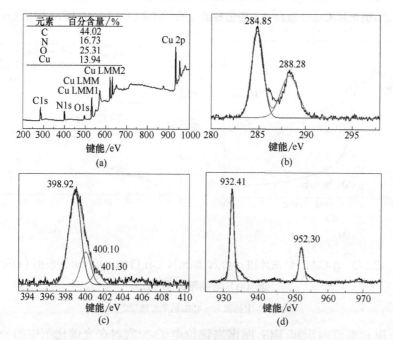

图 3.8　30% g-C_3N_4 纳米片/Cu_2O 复合物的 XPS 谱
（a）总谱（插入表为催化剂的元素分析表）；（b）C1s 谱；（c）N1s 谱；（d）Cu 2p 谱

图 3.8 为光催化剂的 XPS 谱，(a) 图为 30%g-C_3N_4/Cu_2O 的总谱，(b) 图为 C1s 的 XPS 谱图，可以看出，在杂环中存在两个峰，一个是 C=C 对应的 284.85eV，另一个是杂环中 C—N=C 对应的 288.28eV。图 3.8 (c) 为 N1s 谱图，结合能分别为 398.92eV、400.10eV 和 401.30eV[34]，证明了在复合物中 g-C_3N_4 的存在。图 3.8 (d) 为 Cu 2p 在 932.41eV 和 952.30eV 处的峰值，分别对应于 Cu $2p_{3/2}$ 和 Cu $2p_{1/2}$ 的光谱，表明 Cu_2O 存在于光催化剂中。

3.3　光催化降解四环素性能测试

图 3.9 为四环素 (TC) 降解的光催化性能图。可以看出，30% g-C_3N_4 纳米片/Cu_2O 复合物的光催化活性远高于其他样品，光照 100 min 后约有 92.1% 的四环素被分解，这应该归因于复合物的比表面积大，对光诱导电荷-空穴对有更好的分离能力和较快的界面电荷转移速率。然而，随着 g-C_3N_4 纳米片的含量进一步增大，可能导致与 Cu_2O 异质结的构建不完全进而不利于电荷分离。因此，只有当两种催化剂的质量比合适时，才能形成有效的异质结结构。由于 g-C_3N_4 纳米片具有大的比表面积和优良的导电性，与 Cu_2O 形成异质结后可以增加催化剂的比表面积，并且使电子可以快速地转移到 g-C_3N_4 纳米片的表面，进而提高电荷-空穴对的有效分离，增加光催化性能。

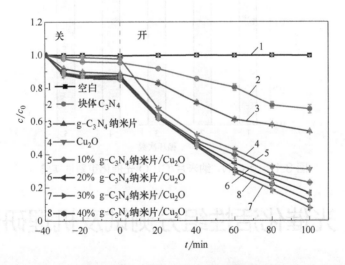

图 3.9　g-C_3N_4纳米片/Cu_2O 复合物光催化降解四环素 (TC) 曲线

通过总有机碳（TOC）分析可以测试四环素溶液的矿化率。如图 3.10 所示，反应 120min 后，有 83.3% 的 TOC 被降解，表明大部分的四环素被 30% g-C_3N_4 纳米片/Cu_2O 复合光催化剂有效矿化降解。

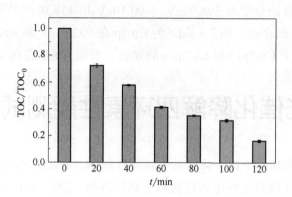

图 3.10　30% g-C_3N_4 纳米片/Cu_2O 复合物在可见光照射下的 TOC

对 g-C_3N_4 纳米片/Cu_2O 复合光催化剂光降解四环素的稳定性进行了研究（图 3.11）。经过 5 次循环，光催化活性仍保持在 88.1%，表明光催化剂具有较高的稳定性。

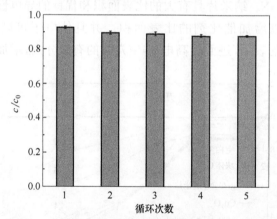

图 3.11　g-C_3N_4 纳米片/Cu_2O 光催化重复性实验

3.4　光催化活性组分测试及机理研究

图 3.12 为加入不同自由基捕获剂的 g-C_3N_4 纳米片/Cu_2O 催化剂对降解

四环素（TC）的影响。添加异丙醇（IPA）后，光反应时间为 100min 时，四环素降解速率没有明显变化，说明 ·OH 应该不是这个过程中的主要活性物种。加入草酸铵（AO）后光催化活性明显降低，说明 h^+ 是参与这一过程的重要活性物种[35, 36]。为了验证电子是否能与被吸附的氧分子发生反应产生超氧负离子自由基（$O_2^-·$），我们通过通入氮气保护来验证，结果可知氮气氛围对降解结果影响也比较显著，说明电子能与氧分子形成超氧负离子自由基（$O_2^-·$）[37]。

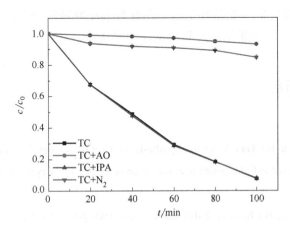

图 3.12　g-C_3N_4 纳米片/Cu_2O 复合光催化剂的活性物种捕获实验

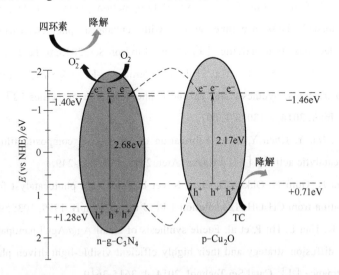

图 3.13　g-C_3N_4 纳米片/Cu_2O 光催化降解四环素机理

g-C_3N_4 纳米片/Cu_2O 光催化降解四环素的机理如图 3.13 所示。当 Cu_2O 与 g-C_3N_4 纳米片/Cu_2O 结合时可以形成 p-n 型异质结结构。在可见光照射下，p 型的 Cu_2O 和 n 型的 g-C_3N_4 都能激发产生电子和空穴对。在内部电场的作用下，Cu_2O 导带上的电子（e^-）可以传送到 g-C_3N_4 导带上[32]，而在 g-C_3N_4 价带上激发态的 h^+ 可以传送到 Cu_2O 价带上[32]。然后电子可以与被吸附的氧分子发生反应，产生超氧负离子自由基（$O_2^{-}·$）进而降解四环素；与此同时，富集于 Cu_2O 导带上的空穴也具有氧化性会迅速与四环素反应。因此 h^+ 和 $O_2^{-}·$是四环素降解的主要活性组分。在这种 p-n 型异质结中，e^- 和 h^+ 均能有效转移和分离，使光催化效率和稳定性明显提高。

参考文献

[1] Zou X, Fan H, Tian Y, et al. Synthesis of Cu_2O/ZnO hetero-nanorod arrays with enhanced visible light-driven photocatalytic activity [J]. Cryst Eng Comm, 2014, 16: 1149-1156.

[2] Xia Y, He Z, Hu K, et al. Fabrication of n-$SrTiO_3$/p-Cu_2O heterojunction composites with enhanced photocatalytic performance [J]. J Alloys Compds, 2018, 753:356-363.

[3] Wang M, Sun L, Lin Z, et al. p-n Heterojunction photoelectrodes composed of Cu_2O-loaded TiO_2 nanotube arrays with enhanced photoelectrochemical and photoelectrocatalytic activities [J]. Energ Environ Sci, 2013, 6: 1211-1220.

[4] Tian Q, Wu W, Sun L, et al. Tube-like ternary α-Fe_2O_3@SnO_2@Cu_2O sandwich heterostructures: synthesis and enhanced photocatalytic properties [J]. ACS Appl Mater Inter, 2014, 6:13088-13097.

[5] Wei S, Ma Y, Chen Y, et al. Fabrication of WO_3/Cu_2O composite films and their photocatalytic activity [J]. J Hazard Ater, 2011, 194: 243-249.

[6] Kakuta N, Goto N, Ohkita H, et al. Silver bromide as a photocatalyst for hydrogen generation from CH_3OH/H_2O solution [J]. J Phys Chem B, 1999, 103: 5917-5919.

[7] Xu Z K, Han L, Hu P, et al. Facile synthesis of small Ag@AgCl nanoparticles via a vapor diffusion strategy and their highly efficient visible-light driven photocatalytic performance [J]. Catal Sci Technol, 2014, 4: 3615-3619.

[8] Tian Y, Chang B, Fu J, et al. Graphitic carbon nitride/Cu_2O heterojunctions: preparat-

ion, characterization, and enhanced photocatalytic activity under visible light [J]. J Solid State Chem, 2014, 212: 1-6.

[9] Liu Y, Dong H, Jia H, et al. CdS nanowires decorated with Cu_2O nanospheres: synthesis, formation process and enhanced photoactivity and stability [J]. J Alloys Compd, 2015, 644: 159-164.

[10] Mousavi M, Habibi-Yangjeh A, Pouran S R. Review on magnetically separable graphitic carbon nitride-based nanocomposites as promising visible-light-driven photocatalysts [J]. J Mater Sci: Mater Electron, 2018, 29: 1719-1747.

[11] Akhundi A, Badiei A, Ziarani G M, et al. Graphitic carbon nitride-based photocatalysts: Toward efficient organic transformation for value-added chemicals production [J]. Mol Catal, 2020, 488: 110902.

[12] Dong H, Guo X, Yang C, et al. Synthesis of $g-C_3N_4$ by different precursors under burning explosion effect and its photocatalytic degradation for tylosin[J]. Appl Catal B, 2018, 230: 65-76.

[13] Wang X, Maeda K, Chen X, et al. Polymer semiconductors for artificial photosynthesis: hydrogen evolution by mesoporous graphitic carbon nitride with visible light[J]. J Am Chem Soc, 2009,131: 1680-1681.

[14] Liu J, Liu Y, Liu N Y, et al, Metal-free efficient photocatalyst for stable visible water splitting via a two-electron pathway [J]. Science, 2015, 347: 970-974.

[15] Zhang X D, Xie X, Wang H, et al, Enhanced photoresponsive ultrathin graphitic-phase C_3N_4 nanosheets for bioimaging [J]. J Am Chem Soc, 2013,135:18-21.

[16] Niu P, Zhang L, Liu G, et al, Graphene-like carbon nitride nanosheets for improved photocatalytic activities [J]. Adv Funct Mater, 2012,22: 4763-4770.

[17] Guo F, Li M Y, Ren H J, et al, Facile bottom-up preparation of cl-doped porous $g-C_3N_4$nanosheets for enhanced photocatalytic degradation of tetracycline under visible light [J]. Sep Purif Technol, 2019, 228:115770-115777.

[18] Vesali-Kermani E, Habibi-Yangjeh A, Ghosh S. Visible-light-induced nitrogen photofixation ability of $g-C_3N_4$ nanosheets decorated with MgO nanoparticles[J]. J Ind Eng Chem, 2020, 84:185-195.

[19] Asadzadeh-Khaneghah S, Habibi-Yangjeh A, Vadivel S. Fabrication of novel $g-C_3N_4$ nanosheet/carbon dots/$Ag_6Si_2O_7$ nanocomposites with high stability and enhanced visible-light photocatalytic activity [J]. J Taiwan Inst Chem Eng, 2019,103: 94-109.

[20] Vesali-Kermani E, Habibi-Yangjeh A, Diarmand-Khalilabad H, et al. Nitrogen photofixation ability of g-C_3N_4 nanosheets/Bi_2MoO_6 heterojunction photocatalyst under visible-light illumination [J]. J Colloid Interf Sci, 2020,563:81-91.

[21] Asadzadeh-Khaneghah S, Habibi-Yangjeh A, Nakata K, Decoration of carbon dots over hydrogen peroxide treated graphitic carbon nitride: exceptional photocatalytic performance in removal of different contaminants under visible light [J]. J Photochem Photobiol A: chem, 2019,374:161-172.

[22] Shao H, Zhao X, Wang Y, et al. Synergetic activation of peroxymonosulfate by Co_3O_4 modified g-C_3N_4 for enhanced degradation of diclofenac sodium under visible light irradiation [J]. Appl Catal B, 2017,218: 810-818.

[23] Hong Y, Li C, Yin B, et al. Promoting visible-light-induced photocatalytic degradation of tetracycline by an efficient and stable beta-Bi_2O_3@g-C_3N_4 core/shell nanocomposite [J]. Chem Eng J, 2018,338: 137-146.

[24] Santosh K, Surendar T, Arabinda B, et al. Synthesis of a novel and stable g-C_3N_4-Ag_3PO_4 hybrid nanocomposite photocatalyst and study of the photocatalytic activity under visible light irradiation [J]. J Mater Chem A, 2013,1: 5333-5340.

[25] Hu K, Li R Q, Ye C L, et al. Facile synthesis of Z-scheme composite of TiO_2 nanorod/g-C_3N_4 nanosheet efficient for photocatalytic degradation of ciprofloxacin [J]. J Cleaner Prod, 2020, 253: 120055.

[26] Li D, Zan J, Wu L, et al. Heterojunction tuning and catalytic efficiency of g-C_3N_4-Cu_2O with glutamate [J]. Ind Eng Chem Res, 2019, 58: 4000-4009.

[27] Liu L, Qi Y, Hu J, et al. Stable Cu_2O@ g-C_3N_4 core@ shell nanostructures: Efficient visible-light photocatalytic hydrogen evolution [J]. Mater Lett, 2015, 158: 278-281.

[28] Wang J, Yang Z, Yao W, et al. Defects modified in the exfoliation of g-C_3N_4 nanosheets via a self-assembly process for improved hydrogen evolution performance [J]. Appl Catal B, 2018, 238: 629-637.

[29] Zhao Q, Wang J L, Li Z P, et al. Heterostructured graphitic-carbon-nitride-nanosheets/copper(I) oxide composite as an enhanced visible light photocatalyst for decomposition of tetracycline antibiotics [J]. Sep Purif Technol, 2020, 250(11), 117238-117245.

[30] Xu J, Zhang L, Shi R, et al. Chemical exfoliation of graphitic carbon nitride for efficient heterogeneous photocatalysis [J]. J Mater Chem A, 2013, 1:14766-14772.

[31] Huang J Y, Hsieh P L, Naresh G, et al. Photocatalytic activity suppression of CdS nanoparticle-decorated Cu_2O octahedra and rhombic dodecahedra [J]. J Phys Chem. C, 2018, 122: 12944-12950.

[32] Wang J, Yang Z, Gao X, et al, Core-shell g-C_3N_4@ ZnO composites as photoanodes with double synergistic effects for enhanced visible-light photoelectrocatalytic activities [J]. Appl Catal B, 2017, 217: 169-180.

[33] Su Y, Ao D, Liu H. MOF-derived yolk-shell CdS microcubes with enhanced visible-light photocatalytic activity and stability for hydrogen evolution [J]. J Mater Chem A, 2017, 5: 8680-8689.

[34] Zou W, Zhang L, Liu L. Engineering the Cu_2O-reduced graphene oxide interface to enhance photocatalytic degradation of organic pollutants under visible light [J]. Appl Catal B, 2016, 181: 495-503.

[35] Shao P, Ren Z, Tian J, et al. Silica hydrogel-mediated dissolution-recrystallization strategy for synthesis of ultrathin α-Fe_2O_3 nanosheets with highly exposed {1 1 0} facets: a superior photocatalyst for degradation of bisphenol S[J]. Chem Eng J, 2017, 323: 64-73.

[36] Shao P, Tian J, Yang F, et al. Identification and regulation of active sites on nanodiamonds: establishing a highly efficient catalytic system for oxidation of organic contaminants [J]. Adv Funct Mater, 2018, 28: 1705295-1705303.

[37] Shao P, Duan X, Xu J, et al. Hetero-geneous activation of peroxymonosulfate by amorphous boron for degradation of bisphenol S [J]. J Hazard Mater, 2017, 322: 532-539.

第 4 章

Z 型异质结光催化剂 UiO-66-NH$_2$/Cu$_2$O

4.1 引言
4.2 UiO-66-NH$_2$/Cu$_2$O 催化剂的表征分析
4.3 光催化降解甲基橙性能测试
4.4 光催化活性组分测试及机理研究

4.1 引言

Z 型异质结是自然界植物光合作用光反应阶段所采用的电荷转移方式。为了模拟自然光合作用，提高光催化剂的氧化还原能力，Bard 在 1979 年提出了传统 Z 型异质结光催化剂结构，由于电子转移过程在图中构成英文字母 Z 的形状，因而称之为 Z 型。Z 型异质结是由 Z-1 和 Z-2 两个错开型的半导体光催化剂和氧化还原电子介体对组成。光激发两个半导体产生电子空穴后，Z-1 的光生空穴 h^+ 同电子给体 D 反应产生电子受体 A，同时 Z-2 的光生电子 e^- 同电子受体 A 反应产生电子给体 D。两种半导体中的电子空穴分别被保留参与氧化还原反应。利用该体系不仅能实现氧化还原位点的空间分离，同时保证了光催化剂能保持合适的价带 1 导带位置，从而保持较强的氧化还原反应能力。

与 p-n 型异质结相比，Z 型异质结光催化剂不仅能有效分离光生电子-空穴，而且具有优异的氧化还原能力。Li 等人[1]通过 Z 型异质结机理合理解释了 Cu_2O/Ag_3PO_4 复合材料的电子-空穴分离，如图 4.1 所示。Ag_3PO_4 导带电位低于 $O_2^{-·}/O_2$ 的标准还原电位，光激发电子不能将 O_2 还原为 $O_2^{-·}$。Ag_3PO_4 价带能级比 $·OH/OH^-$ 和 $·OH/H_2O$ 的更高。Ag_3PO_4 价带中的光生空穴可以与 H_2O 和 OH^- 反应生成羟基自由基（·OH）。对于 Cu_2O/Ag_3PO_4 复合材料，在可见光照射下，Ag_3PO_4 导带处的光致电荷 e^- 将被 Cu_2O 价带处的 h^+ 迅速猝灭。因此，Ag_3PO_4 的价带富含空穴，而 Cu_2O 的导带富含电子，并且

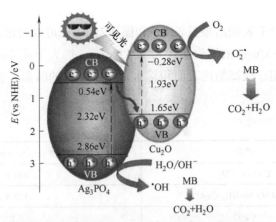

图 4.1 Cu_2O/Ag_3PO_4 的 Z 型异质结机理[1]

可以容易地分离光载流子。因此，制备的 Cu_2O/Ag_3PO_4 复合材料通过 Z 型异质结机理显示出较高的光催化效率。同样，Cu_2O/CuO[2]、$rGO-Cu_2O/Bi_2O_3$[3]、$Cu_2O/Ag/BiVO_4$[4]和 $Cu_2O/Cu/g-C_3N_4$[5]的光催化降解活性也可以用 Z 型异质结机理来解释。

微孔金属有机框架（MOF）由通过有机配体相互连接的金属簇组成[6]。它们具有高比表面积、可调节的孔径和功能化的优点，因此在许多领域引起了广泛关注[7-9]。特别是，许多 MOF 材料可以在光照射下被激发，表现出类似半导体的光催化特性。此外，具有高比表面积的 MOF 材料可以作为理想的光催化剂载体，避免纳米催化剂颗粒的聚集，与其他半导体材料结合可以提高比光催化在废水处理中的活性和稳定性[10,11]。$UiO-66-NH_2$ 因其高比表面积、良好的化学稳定性和具有可见光响应而引起广泛关注。迄今为止，它可以与其他可见光光催化剂如 $BiOBr$[12]、CdS[13] 和 $g-C_3N_4$[14] 等结合，以获得具有优异光催化活性的复合材料[15]。

为了进一步提高 Cu_2O 的光催化活性和稳定性，本文采用简便的浸渍方法制备 $UiO-66-NH_2/Cu_2O$ 复合材料，并将其应用于可见光下的甲基橙降解[16]。

4.2 $UiO-66-NH_2/Cu_2O$ 催化剂的表征分析

催化剂的 BET 表征结果见表 4.1 和图 4.2。从中可以看出，20%（质量分数）$UiO-66-NH_2/Cu_2O$ 的比表面积为 $125.9m^2/g$，远大于纯 Cu_2O（$8.1m^2/g$）。因此，$UiO-66-NH_2$ 较高的比表面积将有利于甲基橙的吸附，提高光催化降解性能。

表 4.1 样品的比表面积

样品	比表面积（m^2/g）
$UiO-66-NH_2$	831.9
$20\%UiO-66-NH_2/Cu_2O$	125.9
Cu_2O	8.1

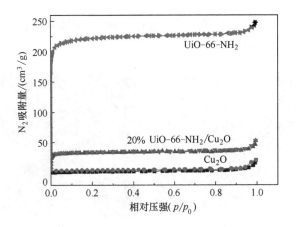

图 4.2　UiO-66-NH$_2$、Cu$_2$O 和 20% UiO-66-NH$_2$/Cu$_2$O 的 N$_2$吸附-解吸等温线

图 4.3 为催化剂的 XRD 图。从图中可以观察到 Cu$_2$O 和 UiO-66-NH$_2$[17]的衍射峰，没有看到 CuO 和 Cu 的衍射峰，这表明 20% UiO-66-NH$_2$/Cu$_2$O 复合材料被成功制备，其中 UiO-66-NH$_2$ 在合成过程中保持其自身的结构。

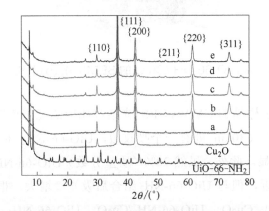

图 4.3　UiO-66-NH$_2$、Cu$_2$O 和不同质量分数 UiO-66-NH$_2$/Cu$_2$O 的 XRD 谱图

质量分数：a—5%；b—10%；c—15%；d—20%；e—25%

图 4.4 给出了 20% UiO-66-NH$_2$/Cu$_2$O 光催化剂的形貌。从图 4.4（a）和（b）可以看出，Cu$_2$O 具有菱方十二面体结构。Cu$_2$O 在 20% UiO-66-NH$_2$/Cu$_2$O 复合材料中的粒径约为 200~300nm[图 4.4（c）]。此外，

观察到 UiO-66-NH$_2$ 均匀地覆盖在 Cu$_2$O 颗粒周围[图 4.4(c)]，表明 Cu$_2$O 和 UiO-66-NH$_2$ 在催化剂中很好地结合。

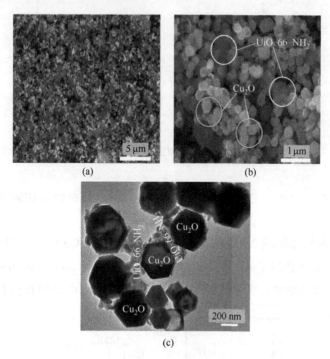

图 4.4 20% UiO-66-NH$_2$/Cu$_2$O 的 SEM[(a), (b)] 和 TEM (c)

图 4.5 给出了样品的 UV-Vis DRS 光谱和相应的能带谱图。如图 4.5(a)所示，Cu$_2$O 在约 630nm 处有一个吸收边，UiO-66-NH$_2$ 也有一个吸收边，对可见光反应明显。此外，虽然 20%（质量分数）UiO-66-NH$_2$/Cu$_2$O 的吸收边缘与 Cu$_2$O 相近，但与 UiO-66-NH$_2$ 结合后吸光度增大。根据图 4.5（b）中的结果计算出 Cu$_2$O、UiO-66-NH$_2$/Cu$_2$O、UiO-66-NH$_2$ 的带隙分别为 1.98eV、1.96eV、2.73eV。因此，UiO-66-NH$_2$ 和 Cu$_2$O 的结合降低了能带带隙，有利于提高对可见光的吸收能力。

光电流和 EIS 用于研究电荷分离效率。众所周知，催化剂中光生 e$^-$-h$^+$ 对的分离效率越高，可测得的光电流就越大。因此，从图 4.6(a)可以看出，20%（质量分数）UiO-66-NH$_2$/Cu$_2$O 复合材料在可见光下表现出更高的电流密度，这意味着该复合材料比纯的 Cu$_2$O 具有更高的光生电子-空穴对分离效

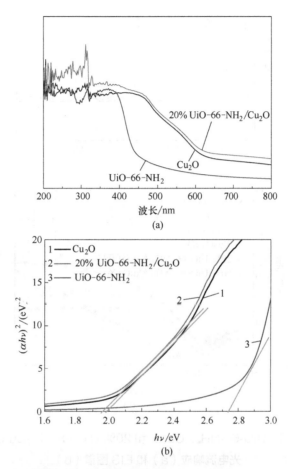

图 4.5 UiO-66-NH$_2$、Cu$_2$O 和 20% UiO-66-NH$_2$/Cu$_2$O 的 UV-vis DRS 光谱(a)和能带谱图（b）

率[18]。图 4.6（b）中的插图是拟合的等效电路图，其中 R_s 是溶液的电阻，R_p 是 20% UiO-66-NH$_2$/Cu$_2$O 与电解质之间界面处的电荷转移电阻，CPE 是常数相元[19]。从图 4.6（b）还可以看出，20% UiO-66-NH$_2$/Cu$_2$O 复合材料的圆半径小于 Cu$_2$O 的圆半径，表明该复合材料具有比 Cu$_2$O 更快的界面电荷-电子转移速率[18]。因此，与纯 Cu$_2$O 相比，20% UiO-66-NH$_2$/Cu$_2$O 复合材料应具有更强的产生和传递光诱导电荷载流子的能力。

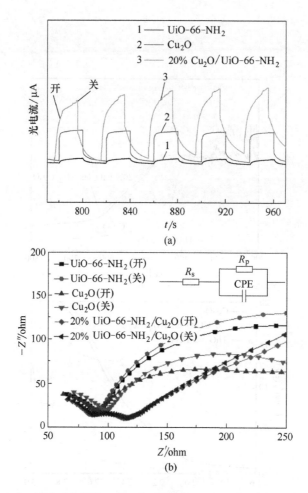

图 4.6 UiO-66-NH$_2$、Cu$_2$O 和 20% UiO-66-NH$_2$/Cu$_2$O 的
光电流响应（a）和 EIS 图谱（b）

插图为拟合的等效电路图

通过 XPS 进一步测量了 20% UiO-66-NH$_2$/Cu$_2$O 复合材料在 5 次循环运行前后的表面化学状态。图 4.7（a）为 XPS 测量光谱，表明 20% UiO-66-NH$_2$/Cu$_2$O 复合材料中存在 Cu、Zr、C 和 O 元素以及各谱线强度在 5 次循环运行前后几乎保持不变。图 4.7（b）为 Zr 3d XPS 光谱。可以看出，Zr 3d 的信号可以分为两个峰值，分别属于 Zr 3d$_{3/2}$（184.5eV）和 Zr 3d$_{5/2}$（182.2eV）[13]。与 UiO-66-NH$_2$ 峰相比，由于与 Cu$_2$O 结合，上述两个峰略有偏移。图 4.7（c）Cu 2p XPS 光谱表明 20% UiO-66-NH$_2$/Cu$_2$O 复

合材料有两个峰,一个在 932.2eV 属于 Cu $2p_{3/2}$ 和另一个是 951.9eV[20] 的 Cu $2p_{1/2}$ 峰。此外,与图中 Cu_2O 的 Cu 2p XPS 光谱相比,在 20% UiO-66-NH_2/Cu_2O 复合材料的废催化剂中未检测到对应于 CuO 中 Cu(Ⅱ) 的 934.0eV 峰,这说明 UiO-66-NH_2/Cu_2O 复合材料中的 Cu_2O 在光降解过程中保持稳定。然而,从图 4.7(d) Cu LMM 光谱中,从废催化剂中检测到痕量 CuO,因为其量太小而无法从 Cu 2p XPS 光谱中观察到。此外,在图 4.7(d) 中未观察到 Cu^0(918.6eV),这意味着在使用的催化剂中未产生 Cu [21]。因此,与 UiO-66-NH_2 结合可以显著抑制 Cu_2O 的光腐蚀,显著提高 Cu_2O 的稳定性,从而获得更高的光催化性能。

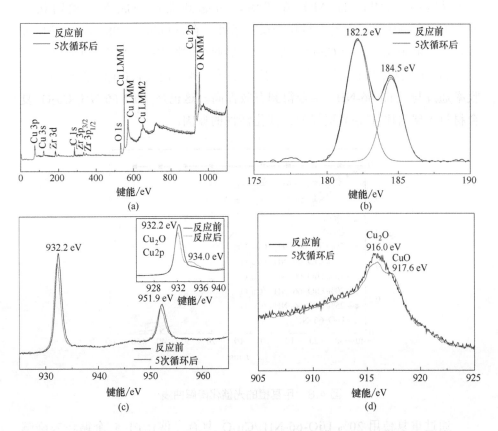

图 4.7 20% UiO-66-NH_2/Cu_2O 复合物的 XPS 光谱

(a) 总谱;(b) Zr 3d 谱;(c) Cu 2p 谱(插图为 Cu_2O 的 Cu 2p 谱);(d) Cu LMM 谱

4.3 光催化降解甲基橙性能测试

通过甲基橙降解研究其光催化性能,得到的结果如图 4.8 所示。图 4.8 中的空白试验表明甲基橙的自光降解可以忽略不计。另外,从图中可见,在黑暗中搅拌 40min,溶液达到吸附平衡后,MO 浓度均明显下降,这是由于该催化剂中 Cu_2O 上的 UiO-66-NH_2 由于较高的比表面积,吸附甲基橙能力强。更重要的是,所有的 UiO-66-NH_2/Cu_2O 复合材料都表现出比纯 Cu_2O 更好的光催化降解率,其中 20%、25% UiO-66-NH_2 Cu_2O(质量分数)的复合材料显示出相近的 MO 降解率,考虑到催化剂成本,我们认为 20%UiO-66-NH_2/Cu_2O 复合材料为最优条件,在 50min 内 98.6% 甲基橙被降解。一般来说,影响 Cu_2O 基催化剂光催化活性的因素很多,包括比表面积、Cu_2O 物种的稳定性、光吸附能力等[22]。如上所述,Cu_2O 的可见光吸收率通过与 UiO-66-NH_2 结合得到有效提高。这也是 UiO-66-NH_2/Cu_2O 复合材料表现出比 Cu_2O 更好的光催化活性的原因。

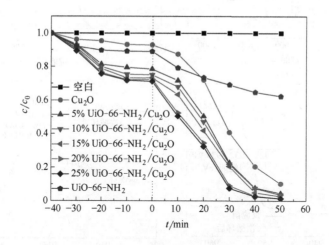

图 4.8 甲基橙的光催化降解曲线

通过重复使用 20% UiO-66-NH_2/Cu_2O 复合光催化剂 5 个循环来降解甲基橙。如图 4.9 所示,即使在第 5 次循环时,该光催化剂的活性仍保持在 90.7%。结果表明,与 UiO-66-NH_2 结合可以显著提高 Cu_2O 光催化剂的稳定性。

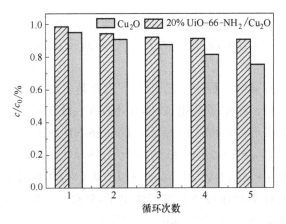

图 4.9　20%UiO-66-NH$_2$/Cu$_2$O（质量分数）光催化重复性实验

4.4　光催化活性组分测试及机理研究

为了探索使用 UiO-66-NH$_2$/Cu$_2$O 复合材料作为催化剂的甲基橙（MO）降解过程中的光催化机理，进行了活性物种捕获实验。从图 4.10 可以看出，加入异丙醇（IPA）（·OH 捕获剂）后，甲基橙的降解率没有明显变化，说明·OH 不是这个过程中的重要贡献者。然而，引入草酸铵（AO）后，光催化活性明显下降，只有少量甲基橙被降解，这意味着空穴是降解过程中的主要物种。此外，UiO-66-NH$_2$/Cu$_2$O 复合材料上的氧分子可以通过光生电子转化为 O$_2^{-}$·。

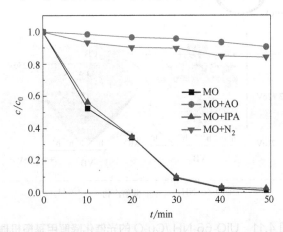

图 4.10　UiO-66-NH$_2$/Cu$_2$O 复合光催化剂的活性物种捕获实验

为了评价 $O_2^{-\cdot}$ 活性物质的影响,还在 N_2 吹扫气氛下进行了催化剂活性评价实验。从图 4.10 可以看出,降解率大大降低,这意味着 $O_2^{-\cdot}$ 也是 UiO-66-NH$_2$/Cu$_2$O 复合材料光降解过程中的主要活性物质。因此,在 UiO-66-NH$_2$/Cu$_2$O 复合催化剂甲基橙降解过程中,h^+ 和 $O_2^{-\cdot}$ 应该是主要的活性物质。

基于上述实验,提出了使用 UiO-66-NH$_2$/Cu$_2$O 复合材料作为催化剂降解甲基橙的可能机理,如图 4.11 所示。UiO-66-NH$_2$/Cu$_2$O 复合催化剂由 p 型 Cu$_2$O 组成 [E_{CB} = -0.23 V 和 E_{VB} = +1.75 V(vs NHE)][22-24]和 n 型 UiO-66-NH$_2$ [E_{VB} = +1.71 V 和 E_{CB} = -1.01 V(vs NHE)][25,26]。因此,在可见光照射下两种半导体可以激发产生电子-空穴对。当 Cu$_2$O 和 UiO-66-NH$_2$ 结合在一起,就可以形成 p-n 型异质结。由于 Cu$_2$O 的导带的电位高于 UiO-66-NH$_2$ 的 CB 电位,UiO-66-NH$_2$ 导带处的光致电荷转移到 Cu$_2$O 的导带上。然而,由于 Cu$_2$O 的导带位于-0.23V(vs NHE),吸附的 O_2 不能接受电子产生 $O_2^{-\cdot}$ [-0.33eV(vs NHE)]。需要说明的是,这与活性物种捕获实验得到的结果不一致,其中 $O_2^{-\cdot}$ 也被认为是光降解过程中的主要活性物种。因此,提出了如图 4.11 所示的 Z 型异质结来解释光催化过程。在此,在可见光照射下,Cu$_2$O 导带处的光致电子将被 UiO-66-NH$_2$ 价带处的 h^+ 快速猝灭。因此,富含 UiO-66-NH$_2$ 的导带的 e^- 可以减少 O_2 产生具有

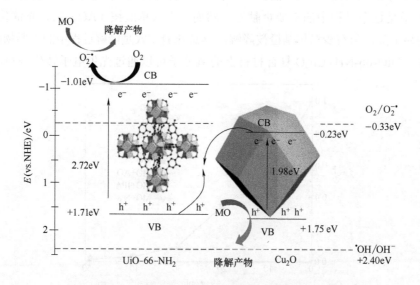

图 4.11 UiO-66-NH$_2$/Cu$_2$O 的光催化降解甲基橙机理

高氧化活性 $O_2^{-}\cdot$ [27]。同时，在活性物质之一的 Cu_2O 上产生的空穴也会与甲基橙快速反应。因此，Z 型异质结可以明显抑制光生电子-空穴对复合。此外，UiO-66-NH_2 和 Cu_2O 的组合降低了带隙，导致更高的可见光吸收能力。因此，光催化效率和稳定性明显提高。

参考文献

［1］ Li Z, Dai K, Zhang J, et al. Facile synthesis of novel octahedral Cu_2O/Ag_3PO_4 composite with enhanced visible light photocatalysis［J］. Mater Lett, 2017, 206: 48-51.

［2］ Wang P, Wang J, Wang X, et al. Cu_2O-rGO-CuO composite: an effective Z-scheme visible-light photocatalyst［J］. Curr Nanosci, 2015, 11: 462-469.

［3］ Shen H, Wang J, Jiang J, et al. All-solid-state Z-scheme system of RGO-Cu_2O/Bi_2O_3 for tetracycline degradation under visible-light irradiation［J］. Chem Eng J, 2017, 313: 508-517.

［4］ Deng Y, Tang L, Zeng G, et al. Plasmonic resonance excited dual Z-scheme $BiVO_4$/Ag/Cu_2O nanocomposite: synthesis and mechanism for enhanced photocatalytic performance in recalcitrant antibiotic degradation［J］. Environ Sci Nano, 2017, 4: 1494-1511.

［5］ Zuo S, Chen Y, Liu W, et al. A facile and novel construction of attapulgite/Cu_2O/Cu/g-C_3N_4 with enhanced photocatalytic activity for antibiotic degradation［J］. Ceram Int, 2017, 43: 3324-3329.

［6］ Hasegawa S, Horike S, Matsuda R, et al. Three-dimensional porous coordination polymer functionalized with amide groups based on tridentate ligand: selective sorption and catalysis［J］. J Am Chem Soc, 2007, 129: 2607-2614.

［7］ Eddaoudi M, Kim J, Rosi N, et al. Systematic design of pore size and functionality in isoreticular MOFs and their application in methane storage［J］. Science, 2002, 295: 469-472.

［8］ Horcajada P, Serre C, Vallet-Regí M, et al. Metal-organic frameworks as efficient materials for drug delivery［J］. Angew Chem, 2006, 45: 5974-5978.

［9］ Ramezanalizadeh H, Manteghi F. Synthesis of a novel MOF/$CuWO_4$ heterostructure for efficient photocatalytic degradation and removal of water pollutants[J]. J Clean Prod, 2018, 172: 2655-2666.

[10] Zhou E H, Li B H, Chen W X, et al. Photocatalytic degradation of organic dyes by a stable and biocompatible Zn (Ⅱ) MOF having ferulic acid: experimental findings and theoretical correlation [J]. J Mol Struct, 2017, 1149: 352-356.

[11] Yang C, You X, Cheng J, et al. A novel visible-light-driven In-based MOF/graphene oxide composite photocatalyst with enhanced photocatalytic activity toward the degradation of amoxicillin [J]. Appl Catal B, 2017, 200: 673-680.

[12] Bibi R, Shen Q, Wei L. Hybrid BiOBr/UiO-66-NH_2 composite with enhanced visible-light driven photocatalytic activity toward RhB dye degradation[J]. RSC Adv, 2018, 8: 2048-2058.

[13] Liang Q, Cui S, Liu C. Construction of CdS@ UIO-66-NH_2 core-shell nanorods for enhanced photocatalytic activity with excellent photostability[J]. J Colloid Interf Sci, 2018, 524: 379-387.

[14] Liang Q, Cui S, Jin J, et al. Fabrication of BiOI@ UIO-66 (NH_2)@g-C_3N_4 ternary Z-scheme heterojunction with enhanced visible-light photocatalytic activity[J]. Appl Surf Sci, 2018, 456: 899-907.

[15] Xu X Y, Chu C, Fu H, et al. Light-responsive UiO-66-NH_2/Ag_3PO_4 MOF-nanoparticle composites for the capture and release of sulfamethoxazole [J]. Chem Eng J, 2018, 350: 436-444.

[16] Zhao Q, Wang J L, Li Z P, et al. UiO-66-NH_2/Cu_2O composite as an enhanced visible light photocatalyst fordeco mposition of organic pollutants [J]. J Photochem Photobiol A: Chem, 2020. 399(8), 112625-112632.

[17] Hao X, Jin Z, Yang H. Peculiar synergetic effect of MoS_2 quantum dots and graphene on Metal-Organic Frameworks for photocatalytic hydrogen evolution[J]. Appl Catal B, 2017, 210: 45-56.

[18] Su Y, Ao D, Liu H. MOF-derived yolk-shell CdS microcubes with enhanced visible-light photocatalytic activity and stability for hydrogen evolution[J]. J Mater Chem A, 2017, 5: 8680-8689.

[19] Sun Q, Peng Y P, Chen H, et al. Photoelectrochemical oxidation of ibuprofen via Cu_2O-doped TiO_2 nanotube arrays [J]. J Hazard Mater, 2016, 319:121-129.

[20] Zou W, Zhang L, Liu L. Engineering the Cu_2O-reduced graphene oxide interface to

enhance photocatalytic degradation of organic pollutants under visible light[J]. Appl Catal B, 2016, 181: 495-503.

[21] Aguilera-Ruiz E, García-Pérez U M, Garza-Galván M, et al. Efficiency of Cu_2O/$BiVO_4$ particles prepared with a new soft procedure on the degradation of dyes under visible-light irradiation [J]. Appl Surf Sci, 2015, 328: 361-367.

[22] Li H, Hong W, Cui Y, et al. Enhancement of the visible light photocatalytic activity of Cu_2O/$BiVO_4$ catalysts synthesized by ultrasonic dispersion method at room temperature [J]. Mater Sci Eng B, 2014, 181:1-8.

[23] Singh M, Jampaiah D, Kandjani A E, et al. Oxygen-deficient photostable Cu_2O for enhanced visible light photocatalytic activity [J]. Nanoscale, 2018, 10: 6039-6050.

[24] Zhang Y, Zhou J, Feng Q, et al. Visible light photocatalytic degradation of MB using UiO-66/g-C_3N_4 heterojunction [J]. Nanocatalyst Chemosphere, 2018, 212: 523-532.

[25] Zhou Y C, Xu X Y, Wang P, et al. Facile fabrication and enhanced photocatalytic performance of visible light responsive UiO-66-NH_2/Ag_2CO_3 composite[J]. Chinese J Catal, 2019, 40: 1912-1923.

[26] Zhou H, Wen Z, Liu J, et al. Z-scheme plasmonic Ag decorated WO_3/Bi_2WO_6 hybrids for enhanced photocatalytic abatement of chlorinated-VOCs under solar light irradiation [J]. Appl Catal B, 2019, 242: 76-84.

[27] Zhao Q, Wang J L, Li Z P, et al. Preparation of Cu_2O/exfoliated graphite composites with high visible light photocatalytic performance and stability[J]. Ceram Int, 2016, 42: 13273-13277.

第 5 章

肖特基型异质结光催化剂 Ti₃C₂Tx/Cu₂O

- 5.1 引言
- 5.2 Ti₃C₂Tx/Cu₂O 催化剂的结构与性能表征
- 5.3 光催化降解四环素性能测试
- 5.4 光催化机理研究

5.1 引言

肖特基型异质结的催化剂是多种异质结催化剂中的一种，是由金属和半导体形成的异质结构，比较明显的特征是其具有整流性质，可以对电流形态进行调整。Schottky[1]在1938年较系统地研究了金属-半导体接触，并指出在界面附近的半导体一侧存在一个势垒，并近似地确定了势垒的形状和电流输运机制。以n型半导体TiO_2为例，金属和半导体接触时，电子从更高费米能级的TiO_2迁移到沉积金属上，直到两个能级匹配，形成肖特基势垒，有效抑制电子和空穴的复合。这导致金属有过多的负电荷，而半导体有过多的正电荷，在电荷耗尽层保持了电荷分离。随着空穴在TiO_2价带的积累，肖特基型异质结在界面形成，迫使光生电子远离空穴。而且光激发肖特基型异质结时，从半导体越过界面进入金属的光致电子并不发生积累，而是直接形成飘移电流流走。因此，肖特基势垒的形成致使光致电荷分离，快速连续迁移分离的光致电子，有利于空穴参与反应，提高光催化活性。Zhang等人[2]解释了Au/Cu_2O的光催化机理，如图5.1所示。当Au与Cu_2O结合时，由于在界面处形成肖特基型异质结，电子将从Au迁移到Cu_2O以平衡费米能级。在可见光照射下，Cu_2O导带中的光生电子会快速移动到Au纳米粒子，而空穴停留在Cu_2O的价带上。因此，光生电子和空穴被有效地分离。分离

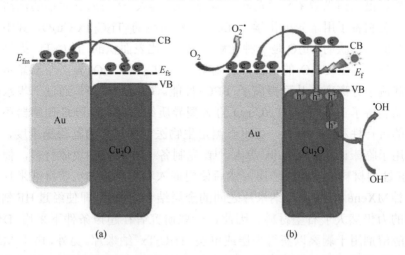

图5.1 Au/Cu_2O的肖特基型异质结机理[2]

(a) 电子迁移方式一；(b) 电子迁移方式二

的电子被吸附的 O_2 俘获产生 $O_2^{-\cdot}$；分离的空穴被表面 OH^- 俘获产生 $\cdot OH$。因此，有机污染物被两个自由基有效降解。此外，Cu/Cu_2O[3]、Ag/Cu_2O[4]和 Zn/Cu_2O[5]的光催化机理也可由上述机理解释。

Ti_3C_2Tx 是一种新型的催化 MXene 材料，具有良好的金属性、亲水性和丰富的表面活性位点。其组成形式上可以分为两类，一类分子组成为 $M_{n+1}X_n$，另一类为 $M_{n+1}X_nTx$，其中 M 为过渡金属(如 Sc、Ti、V、Cr、Zr、Hf、Nb、Mo、Ta 和 W)，X 则代表 C 或 N，Tx 是终端官能团（如 OH、O 或 F)，而 n 通常是一个 1~3 的整数。这种独特的结构是由原始的块状 MAX 相（$M_{n+1}AX_n$），经过选择性地侵蚀反应最活跃的组分 A （ⅢA 或ⅣA 元素）而得到的。MXene 被侵蚀后具有片状结构，其横向尺寸大于 100nm，或高达几微米甚至更大，但厚度通常小于 5nm [6-9]。通过将体积材料的厚度减小至原子厚度，使配位数、键长、键角、原子无序度等局部原子排列发生变化，其性质也会产生巨大的变化[10]。由于其独特的金属性，它可以与其他半导体结合形成肖特基型异质结。

迄今为止，肖特基型异质结的构建策略已被广泛研究来提高染料降解[11]、制氢[12-18]和 CO_2 还原[19-22]等领域的光催化剂性能。例如，制备的 Ti_3C_2Tx/Cu_2O 异质结表现出高化学稳定性、低毒性和高氧化能力 [23-25]。Fu 等人 [23]通过简便的浸涂法制备了用于分解水的 Ti_3C_2Tx/Cu_2O 光催化剂，发现 Ti_3C_2Tx/Cu_2O 的优异性能应归功于 Ti_3C_2Tx 中富集的氧空位，它可以增强导电性、提高捕光效率和电荷转移能力。Li 等人 [24]还通过简单的油浴加热合成工艺制备了用于光电化学（PEC）传感系统的 Ti_3C_2Tx/Cu_2O，其中具有规则八面体形状的 Cu_2O 是一种光敏材料，大比表面积的 Ti_3C_2Tx 提供了许多活性位点用于 Cu_2O 的生长。所制备的 Ti_3C_2Tx/Cu_2O 异质结构显示出较高的载流子分离率，从而提高了 PEC 性能。Chen 等人 [25]通过一步水热反应成功制备了 $TiO_2/Ti_3C_2Tx/Cu_2O$ 的 Z 型异质结。制备的 PEC 传感器还具有良好的可行性和高选择性，可用于测定生物医学样品中的葡萄糖浓度，表明其可用于临床诊断。Naguib 等人[26]首先制备了 Ti_3C_2Tx 块体材料，他们使用 HF 蚀刻材料中的铝原子层，然后使用嵌入和剥离的方法获得纳米片。对于这种 MXene 材料，由于纳米片之间的金属结合力很强，即使经过 HF 刻蚀，层间的力仍然大于石墨材料。因此，一些研究者在超声条件下采用 DMSO 或乙醇溶剂用于剥离以获得少层或单层 Ti_3C_2Tx 纳米片。另外，由于 MXene 是类石墨材料，因此可以通过 Hummers 方法进行膨胀和剥离。然而前期还没有关于用这种方法制备 Ti_3C_2Tx 纳米片的报道。所以在本研究中，我们采用膨胀法制备了比表面积大的 Ti_3C_2Tx 纳米片，然后用简单的共沉淀法制备了 Ti_3C_2Tx/Cu_2O 复合材料，并将其应用于可见光下对四环素的光催化降解。根据实验结果，提出了可能的光催化机理[27]。

5.2 Ti_3C_2Tx/Cu_2O 催化剂的结构与性能表征

表 5.1 和图 5.2 为催化剂的 BET 表征和对比。可以看出，通过膨胀法获得的 Ti_3C_2Tx 纳米片的比表面积比 Ti_3C_2Tx 高约 7 倍。此外，与其他方法获得的 Ti_3C_2Tx 纳米片相比，采用这种膨胀法制备的 Ti_3C_2Tx 具有更高的比表面积，有利于四环素的吸附，从而提高光催化脱除效率。

表 5.1 膨胀法和其他方法制备 Ti_3C_2Tx 样品的比表面积对比

样品	制备方法	比表面积/(m^2/g)	参考文献
Ti_3C_2Tx	二甲基亚砜(超声法)	23.00	[28]
Ti_3C_2Tx	乙醇(超声法)	13.00	[29]
Ti_3C_2Tx	表面活性剂	22.50	[30]
Ti_3C_2Tx	水热法	44.60	[31]
Ti_3C_2Tx	膨胀法	25.38	本研究
Ti_3C_2Tx 块体		3.70	本研究

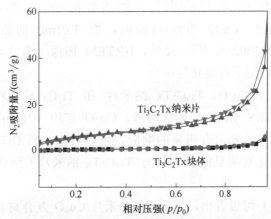

图 5.2 Ti_3C_2Tx 块体和 Ti_3C_2Tx 纳米片的 N_2 吸附-解吸等温线

图 5.3 显示了 Ti₃C₂Tx 纳米片、Cu₂O 和 Ti₃C₂Tx-纳米片/Cu₂O 复合材料的 XRD 光谱。可以看到与 Ti₃AlC₂ 的{104}面对应的 39°的峰消失了，表明使用本方法有效去除了层间的 Al 原子。同时，观察到对应 8.8°的 Ti₃C₂Tx 纳米片的峰值，表明使用膨胀方法成功形成了 Ti₃C₂Tx 片[28]。此外，29.3°、36.2°、42.1°、61.3°和 73.4°处的峰对应于 Cu₂O 的{110}、{111}、{200}、{220}和{311}面[29]。在 Ti₃C₂Tx 纳米片/Cu₂O 复合材料中也发现了 Ti₃C₂Tx 纳米片在 8.8°处的衍射峰，表明复合材料中 Ti₃C₂Tx 纳米片与 Cu₂O 成功结合。

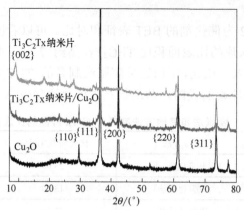

图 5.3 Ti₃C₂Tx 纳米片、Cu₂O 和 Ti₃C₂Tx 纳米片/Cu₂O 的 XRD 光谱

从图 5.4（a）中可以观察到 Ti₃C₂Tx 纳米片为单层或多层结构。从图 5.4（c）可以看出，Cu₂O 具有立方体结构，尺寸约为 80~120nm，而 Ti₃C₂Tx 纳米片分散并包围在 Cu₂O 颗粒周围。从图 5.4（b）和（e）可以看出，0.246nm 的晶面间距对应于 Cu₂O 的{111}晶面，而 1.01nm 的晶面间距对应于 Ti₃C₂Tx 纳米片的{002}面[29]。此外，HRTEM 图像[图 5.4（d）]证明了 Cu₂O 和 Ti₃C₂Tx 纳米片成功的结合。

图 5.5 显示了 Cu₂O、Ti₃C₂Tx 纳米片 和 Ti₃C₂Tx 纳米片/Cu₂O 复合材料的 UV-Vis DRS 光谱。如图所示，Cu₂O 在约 620nm 处有一个吸收边，Ti₃C₂Tx 对可见光也有明显的响应。可以看出 Ti₃C₂Tx/纳米片/Cu₂O 复合材料对可见光有明显的响应，与 Ti₃C₂Tx 纳米片的结合进一步增强了吸收强度。

从图 5.6（a）可以看出，Ti₃C₂Tx 纳米片/Cu₂O 复合材料的光电流值明显高于纯 Cu₂O，表明 Ti₃C₂Tx 的电子-空穴分离更有效。图 5.6（b）显示了样品的 EIS 图。可以看到 Ti_3C_2Tx 纳米片/Cu_2O 复合材料的圆半径小于 Cu_2O，表明该复合材料的界面电荷电子转移速率比 Cu_2O 快[30]。因此，与纯 Cu_2O 相比，Ti_3C_2Tx 纳米片/Cu_2O 复合材料应具有更强的分离和转移光致电荷载流子的能力。

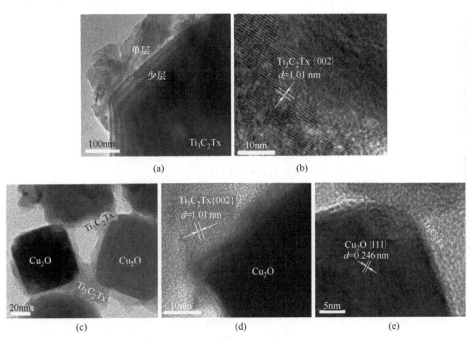

图 5.4　Ti_3C_2Tx 纳米片的 TEM(a)和 HRTEM (b)，Ti_3C_2Tx 纳米片/Cu_2O 复合物的 TEM（c）和 HRTEM 图像 [(d), (e)]

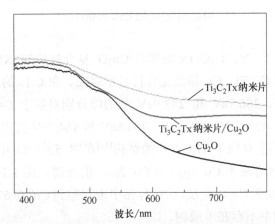

图 5.5　Cu_2O、Ti_3C_2Tx 纳米片和 Ti_3C_2Tx 纳米片/Cu_2O 的 UV-Vis DRS 光谱

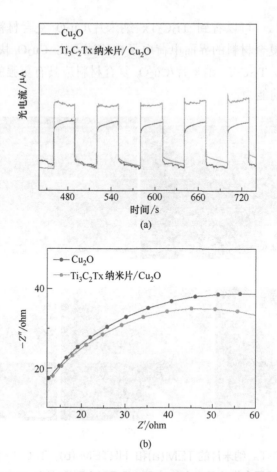

图 5.6 Cu₂O 和 Ti₃C₂Tx 纳米片/Cu₂O 复合物的光
电流响应(a)和 EIS 图谱(b)

图 5.7 显示了 7% Ti₃C₂Tx 纳米片/Cu₂O 复合材料的 XPS 分析结果,如图所示,元素 C、Ti、Cu 都在复合材料中存在。在 C 1s 的光谱中[图 5.7(b)],284.8eV、286.7eV 和 288.9eV 处的峰分别对应于 C═C、C—O 和 C—F。同时,在 455.7eV、458.5eV、460.8eV 和 464.2eV 处出现峰值的 Ti 2p 光谱[图 5.7(c)]对应于 Ti₃C₂Tx 的结构[31]在图 5.6(d)中,932.46eV 和 952.20eV 处的峰对应于 Cu $2p_{3/2}$ 和 Cu $2p_{1/2}$ 的光谱。图 5.7(d)中出现在 570.0eV 处的峰表明 Cu₂O 存在,而在样品中未观察到 Cu⁰ 在 571.2eV 处的峰,这表明光催化剂中不存在单质铜。

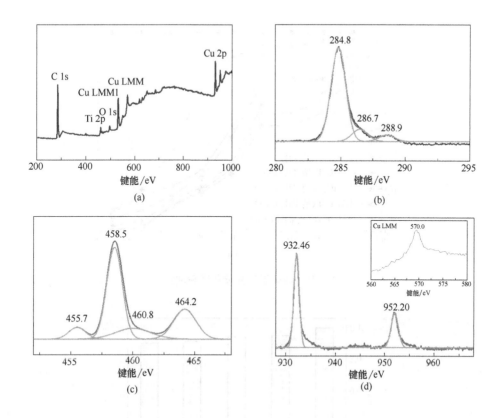

图 5.7　7%Ti₃C₂Tx 纳米片/Cu₂O 复合物的 XPS 光谱

（a）总谱；（b）C 1s 谱；（c）Ti 2p 谱；（d）Cu 2p 谱（插图为 Cu LMM 谱）

5.3　光催化降解四环素性能测试

从图 5.8 可以看出，Ti_3C_2Tx 纳米片/Cu_2O 复合材料的光催化降解率明显高于纯 Cu_2O，其中 7% Ti_3C_2Tx 纳米片/Cu_2O 复合材料的四环素降解性能最好，50min 内光催化降解率为 97.6%。Ti_3C_2Tx 纳米片具有优异的导电性，与 Cu_2O 结合后可以增加催化剂的比表面积并提供更多的活性位点，特别是使电子能够快速转移到 Cu_2O 表面，从而促进光生载流子的分离和迁移，以提高光催化活性。如图 5.9 所示，对 7% Ti_3C_2Tx 纳米片/Cu_2O 复合材料光催化降解四环素的稳定性进行了研究，结果显示经过 5 个循环，光催化活性仍保持在 90.1%，表明光催化剂具有较高的稳定性。

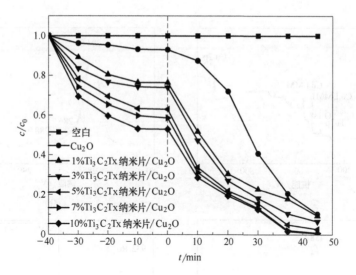

图 5.8 四环素的光催化降解曲线

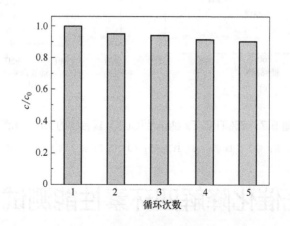

图 5.9 Ti_3C_2Tx 纳米片/Cu_2O 光催化重复性实验

5.4 光催化机理研究

图 5.10 显示了使用 7% Ti_3C_2Tx 纳米片/Cu_2O 复合材料作为光催化剂的活性物种捕获实验的结果。可以看到在引入草酸铵（AO）（h^+ 捕获剂）和 N_2 吹扫下，光催化活性降低不明显，只有少量四环素被降解，表明 h^+ 和 $O_2^-\cdot$ 是四环素降解的主要物种。

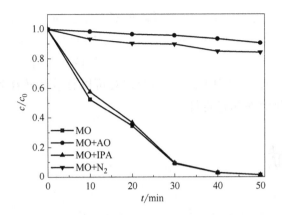

图 5.10　7% Ti₃C₂Tx 纳米片/Cu₂O 复合光催化剂
（质量分数）的活性物种捕获实验

光催化降解四环素的机理如图 5.11 所示。当 Cu₂O 与 Ti₃C₂Tx 纳米片结合时，会形成肖特基异质结。在可见光照射下，电子被激发并从价带跃迁到 Cu₂O 的导带，在价带中产生空穴。由于 Ti₃C₂Tx 具有类金属特性，Ti₃C₂Tx 与 p 型半导体 Cu₂O 之间可以形成肖特基势垒，进一步促进电子从 Cu₂O 向 Ti₃C₂Tx 表面单向流动，从而有效促进光生电子的分离和传输[32]，同时，迁

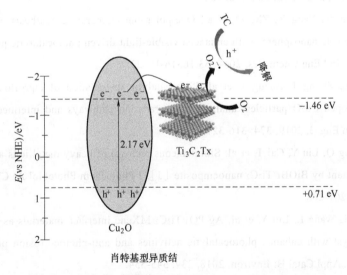

图 5.11　Ti₃C₂Tx 纳米片/Cu₂O 的光催化降解四环素机理

移到 Ti_3C_2Tx 表面的电子可以与吸附的氧分子反应生成超氧负离子自由基（$O_2^-·$）。Cu_2O 上富集的 h^+ 也会与四环素快速反应。因此，催化剂表面的（$O_2^-·$ 和 h^+）与四环素分子发生反应，导致四环素降解和矿化[32]。在肖特基异质结中，该催化剂中的电子-空穴对可以有效地转移和分离，从而显著提高了光催化效率和稳定性。

参考文献

[1] Schottky W. Halbleitertheorie der sperrschicht [J]. Naturwissenschaften, 1938, 26 (52): 843.

[2] Zhang W, Wang B, Hao C, et al. Au/Cu₂O Schottky contact heterostructures with enhanced photocatalytic activity in dye decomposition and photoelectrochemical water splitting under visible light irradiation [J]. J Alloys Compds, 2016, 684: 445-452.

[3] Ai Z, Zhang L, Lee S, et al. Interfacial hydrothermal synthesis of Cu@ Cu₂O core shell microspheres with enhanced visible-light-driven photocatalytic activity [J]. J Phys Chem C, 2009, 113: 20896-20902.

[4] Zhang W, Yang X, Zhu Q, et al. One-pot room temperature synthesis of Cu₂O/Ag composite nanospheres with enhanced visible-light-driven photocatalytic performance [J]. Ind Eng Chem Res, 2014, 53:16316-16323.

[5] Yu X, Zhang J, Zhang J, et al. Photocatalytic degradation of ciprofloxacin using Zn-doped Cu₂O particles: analysis of degradation pathways and intermediates [J]. Chem Eng. J, 2019, 374: 316-327.

[6] Huang Q, Liu Y, Cai T, et al. Simultaneous removal of heavy metal ions and organic pollutant by BiOBr/Ti₃C₂ nanocmposite [J]. J Photochem Photobiol A: Chem, 2019, 375: 201-208.

[7] Cai T, Wang L, Liu Y, et al. Ag₃PO₄/Ti₃C₂ MXene interface materials as a Schottky catalyst with enhance photocatalytic activities and anti-photocorrosion performance [J]. Appl Catal B: Environ, 2018, 239: 545-554.

[8] Liu N, Lu N, Su Y, et al. Fabrication of Ti₃C₂/g-C₃N₄ composite and its visible-light photocatalytic capability for ciprofloxacin degradation[J]. Sep Purif Tech, 2019, 211:

782-789.

[9] Fang Y, Cao Y, Chen Q. Synthesis of an Ag_2WO_4/Ti_3C_2 Schottky composite by electrostatic traction and its photocatalytic activity [J]. Ceramics Int, 2019, 45: 22298-22307.

[10] Zhou W, Zhu J, Wang F, et al. One-step synthesis of Ceria/Ti_3C_2 nanocomposites with enhanced photocatalytic activity [J]. Mater Lett, 2017, 206: 237-240.

[11] Wang H, Sun Y, Wu Y, et al. Electrical promotion of spatially photoinduced charge separation via interfacial-built-in quasi-alloying effect in hierarchical $Zn_2In_2S_5$/$Ti_3C_2(O,OH)_x$ hybrids toward efficientphotocatalytic hydrogen evolution and environmental remediation [J]. Appl Catal B Environ, 2019, 245: 290-301.

[12] Li Y, Deng X, Tian J, et al. Ti_3C_2 MXene-derived Ti_3C_2/TiO_2 nanoflowers for noble-metal-free photocatalytic overall water splitting [J]. Appl Mater Today, 2018, 13: 217-227.

[13] Lin P, Shen J, Yu X, et al. Construction of Ti_3C_2 MXene/O-doped g-C_3N_4 2D/2D Schottky-junction for enhanced photocatalytic hydrogen evolution [J]. Ceram Int, 2019, 45: 24656-24663.

[14] Tian P, He X, Zhao L, et al. Enhanced charge transfer for efficient photocatalytic H_2 evolution over UiO-66-NH_2 with annealed Ti_3C_2Tx MXenes [J]. Int J Hydrogen Energy, 2019, 44:788-800

[15] Li Y, Yin Z, Ji G, et al. 2D/2D/2D heterojunction of Ti_3C_2 MXene/MoS_2 nanosheets/ TiO_2 nanosheets with exposed {001} facets toward enhanced photocatalytic hydrogen production activity [J]. Appl Catal B: Environ, 2019, 246; 12-20.

[16] Li Z, Zhang H, Wang L, et al. 2D/2D BiOBr/Ti_3C_2 heterojunction with dual applications in both water detoxification and water splitting[J]. J Photochem Photobiol: Chem, 2020, 386: 112099-112107.

[17] An X, Wang W, Wang J, et al. The synergetic effect of Ti_3C_2 MXene and Pt as co-catalysts for highly efficient photocatalytic hydrogen evolution over g-C_3N_4 [J]. Phys Chem Chem Phys, 2018, 20: 11405-11411.

[18] Xiao R, Zhao C, Zou Z, et al. In situ fabrication of 1D CdS nanorod/2D Ti_3C_2 MXene nanosheet Schottky heterojunction toward enhanced photocatalytic hydrogen evoluti-on [J]. Appl Catal B: Environ, 2019, 250: 118382.

[19] Low J, Zhang L, Tong T, et al. TiO_2/MXene Ti_3C_2 composite with excellent photocata

lytic CO_2 reduction activity [J]. J Catal, 2018, 361: 255-266.

[20] Shen J, Shen J, Zhang, et al. Built-in electric field induced CeO_2/Ti_3C_2 MXene Schottky-junction for coupled photocatalytic tetracycline degradation and CO_2 reduction [J]. Ceram Int, 2019, 45 : 24146-24153.

[21] Cao S, Shen B, Tong T, et al. 2D/2D heterojunction of ultrathin $MXene/Bi_2WO_6$ nanosheets for improved photocatalytic CO_2 reduction[J] Adv Funct Mater, 2018, 28: 1800136.

[22] Ye M, Wang X, Liu E, et al. Boosting the photocatalytic activity of P25 for carbon dioxide reduction by using a surface-alkalinized titanium carbide MXene as cocatalyst [J]. ChemSusChem, 2018, 11: 1606-1611.

[23] Fu X C, Hui C A, Shang Z C, et al. Three-dimensional Cu_2O nanorods modified by hydrogen treated Ti_3C_2Tx mxene with enriched oxygen vacancies as a photocathode and a tandem cell for unassisted solar water splitting [J]. Chem Eng J, 2020, 381: 122001-122011.

[24] Li M, Wang H, Wang X, et al. Ti_3C_2/Cu_2O heterostructure based signal-off phot-oelectrochemical sensor for high sensitivity detection of glucose [J]. Biosens Bioelectron, 2019, 142: 111535-111541.

[25] Chen G, Wang H, Wei X, et al. Efficient Z-scheme heterostructure based on $TiO_2/Ti_3C_2/Cu_2O$ to boost photoelectrochemical response for ultrasensitive biosensing [J]. Sensor Actuat B-Chem, 2020, 312: 127951.

[26] Naguib M, Kurtoglu M, Presser V, et al. Two-dimensional nanocrystals produced by exfoliation of Ti_3AlC_2 [J]. Adv Mater, 2011, 23: 4248-4253.

[27] Zhao Q, Wang J L, Li Z P, et al. Two-dimensional Ti_3C_2Tx-nanosheets/Cu_2O composite as a high-performance photocatalyst for decomposition of tetracycline [J]. Carbon Resources Conversion, 2021, 4: 197-204.

[28] Wu F, Luo K, Huang C, et al. Theoretical understanding of magnetic and electronic structures of Ti_3C_2 monolayer and its derivatives[J]. Solid State Commun, 2015, 222: 9-13.

[29] Hantanasirisakul K, Zhao M Q, Urbankowski P, et al. Fabrication of Ti_3C_2Tx MXene transparent thin films with tunable optoelectronic properties[J]. Adv Electron Mater, 2016, 2: 1600050-1600057.

[30] Wen Y, Rufford T E, Chen X, et al. Ti_3C_2Tx Nitrogen-doped Ti_3C_2Tx MXene

electrodes for high performance super capacitors [J]. Nano Energy, 2017, 38: 368-376.

[31] Lin P, Shen J, Yu X, et al. Construction of Ti_3C_2Tx/O-doped g-C_3N_4 2D/2D Schottky-junction for enhanced photocatalytic hydrogen evolution [J]. Ceram Int, 2019, 45: 24656-24663.

[32] Wei Z, XinT, Xiao W, et al. Novel pn heterojunction photocatalyst fabricated by flower-like $BiVO_4$ and Ag_2S nanoparticles: simple synthesis and excellent photocatalytic performance [J]. Chem Eng J, 2019, 361:1173-1181.

第 6 章

Cu$_2$O 复合其他材料光催化剂

6.1 引言
6.2 KAPs-B/Cu$_2$O 光催化剂光催化性能研究
6.3 Cu$_2$O 复合碳材料的研究

6.1 引言

为了进一步提高 Cu_2O 基复合光催化剂的催化性能,本章进一步对 Cu_2O 复合其他材料光催化剂的催化性能进行了深入研究。

超交联聚合物(HCPs)是一系列永久微孔聚合物材料,最初由 Tsyurupa 发现,HCPs 凭借其显著的优点快速发展并已获得越来越高的关注[1]。HCPs 的合成主要基于傅-克反应,该反应表现为自身的快速动力学,可以形成强的键合作用,使单体形成具有高孔隙率的高度交联的网络。由于合成方法简单且通用,大量的芳香族单体可用于开发具有各种孔结构的聚合物网络,或可利用特定官能团提高其比表面积。此外,HCPs 的常规合成方法只需低成本试剂(单体、反应介质和催化剂),并易于处理和控制反应条件生产高产率产物。HCPs 已被广泛用于废水处理、有机蒸汽吸附等[2]。特别是,HCPs 具有丰富的微孔,可以让被吸附物分子从表面转移到内部微孔,从而大大提高吸附能力和吸附速率。此外,由于其疏水骨架,HCPs 对有机分子具有很强的亲和力[3]。最近,一系列基于 HCPs 的催化剂被报道用于有机转化[4-7]。然而,尚未见 HCPs 材料应用于可见光降解有机污染物的光催化剂的报道。

另外,碳材料因其作为催化剂载体和电极材料等在催化领域有着潜在的应用而得到了广泛研究。研究已表明将碳材料与半导体光催化剂相结合是提高其光催化性能的有效策略。

(1) Cu_2O/碳球(CS)

在碳材料的各种结构形式中,碳球(CS)以其优异的物理和化学性质在不同领域引起了广泛研究。Zhou 等人[8] 使用醋酸铜作为前驱体,碳球作为载体材料,制备了 Cu_2O/CS 异质结构。从图 6.1 可见,其中 Cu_2O 纳米球通过相对简单的化学还原方法沉积在碳球表面。通过甲基橙降解评价其光催化性能,与纯 Cu_2O 相比,Cu_2O/CS 复合材

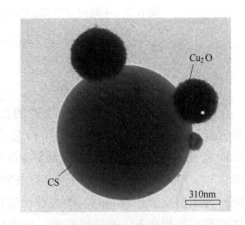

图 6.1 Cu_2O/CS 结构的 SEM 图像[8]

料的光催化活性有所提高，这归因于复合材料的良好吸附性能以及 Cu_2O 和碳球之间光生电子-空穴对的分离。

（2）Cu_2O/碳纳米管

近年来一些研究表明，Cu_2O 纳米颗粒与碳纳米管（CNTs）的结合可以显著提高光催化性能，因为碳纳米管促进了 Cu_2O 纳米颗粒的分散和电子传输[9]。Cu_2O/CNTs 复合材料已通过多种方法制备[10]，如图 6.2 所示，Yang 等人[11]首次通过简便的湿化学方法制备了一种新型的 Cu_2O/CNTs 分级花状纳米复合材料（图 6.2），该复合材料由平均直径约为 80~120nm 的花状结构组成。通过在可见光照射下降解苯酚来评价其催化性能，与其他制备的 Cu_2O/CNTs 相比，Cu_2O/CNTs 花状纳米复合材料表现出优异的光催化性能，这是因为这种结构扩大了吸收光的范围，改善了 Cu_2O 纳米颗粒与 CNTs 的连接并增强了可见光吸收强度。

图 6.2　不同尺度下 Cu_2O/CNTs 结构的 SEM 图像[11]

(a) 5μm；(b) 500nm

（3）Cu_2O/ 膨胀石墨（EG）

由于膨胀石墨（EG）具有独特的二维共轭结构，可以为光催化反应提供三维环境（图 6.3），通过与 Cu_2O 复合，可以快速分离光生电荷对，进而降低电子对-空穴的复合率。因此与 EG 结合可以显著提高 Cu_2O 的光催化降解效率。Zhao 等人[12]首先通过沉淀法制备了 Cu_2O/ EG，并以甲基橙溶液的降解为模型反应评价该材料的光催化活性。在复合材料中，Cu_2O 以纳米晶体的形式沉积在 EG 的片层表面。与纯 Cu_2O 相比，Cu_2O/ EG 复合材料的光催化性能明显增强，甲基橙降解 60min 的效率可达 96.7%。

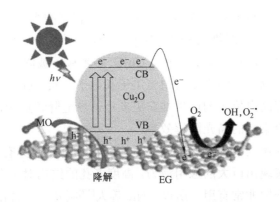

图 6.3 Cu$_2$O/EG 复合材料光催化降解机理[12]

（4）Cu$_2$O/氧化石墨烯（GO）

氧化石墨烯（GO）是石墨烯最重要的衍生物之一，其特征是具有层状结构，其基面和边缘带有氧官能团。由于以下几个方面，GO 吸引了研究人员的最多关注。首先，由于存在局部 sp^2 杂化结构，GO 可以被认为是一种半导体材料。如果将 Cu$_2$O 粒子组装在 GO 表面，Cu$_2$O 中诱导的光生电子可以很容易地迁移到 GO，从而避免了 Cu$_2$O 中光生电子-空穴对的表面复合[13]。图 6.4 为制备的 Cu$_2$O/GO 结构的 SEM 图，从图中可以看到，Cu$_2$O 颗粒均匀地分散在 GO 表面。其次，GO 纳米片表面有许多含氧基团，有利

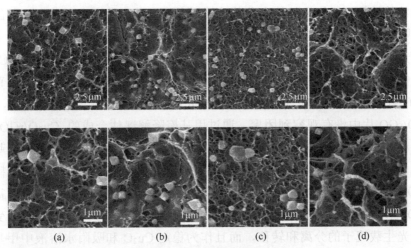

图 6.4 Cu$_2$O/GO 结构的 SEM 图像[13]

（a）~（d）为不同视角下的图像

于有机染料吸附在 GO 表面,从而提高复合光催化剂的光催化降解效率[14,15]。Tu 等人[16]通过在再生纤维素(RC)/GO 复合膜的微孔中原位合成 Cu_2O 制备了可见光光催化剂。在复合材料中,Cu_2O 纳米粒子被固定并均匀分布在再生纤维素基体中以激发和产生自由光电子和电子空穴,GO 片转移产生的光电子以防止局部高电位区的产生并在更多点诱导链降解反应,导致在可见光照射下对甲基橙染料的高光降解效率。

泡沫属于众所周知的多孔材料,具有极低的密度但具有很高的孔隙率,在各个领域都表现出巨大的应用潜力。高度多孔的结构对于有机染料的渗透和光催化剂的降解非常有用。所以,Nie 等人[17]制备了一种由再生微晶纤维素(MCC)纤维和 Cu_2O/GO 复合光催化剂组成的新型复合泡沫,并通过亚甲基蓝降解测量其光催化性能。复合泡沫表现出典型的多孔结构,此外,由于 Cu_2O 和 GO 之间的氢键相互作用,促进了 Cu_2O 在复合泡沫中的分散。MCC/Cu_2O/GO 复合泡沫对亚甲基蓝表现出优异的光催化降解能力,远高于使用其他 Cu_2O 基光催化降解材料的文献报道。这主要归因于泡沫的多孔结构和 Cu_2O/GO 复合结构的形成,显示了"微反应器"在亚甲基蓝光催化降解过程中的作用。

(5) Cu_2O/rGO

还原氧化石墨烯(rGO)作为一种具有应用前景良好的材料,具有一系列独特的性能,例如卓越的电荷载流子迁移率、极高的比表面积和优异的导电性[18],近年来受到了人们的极大关注。它还用作柔性导电载体,在许多领域中封装 Cu_2O 纳米晶体。Abulizi 等人[19]使用简单的声化学途径制备了 Cu_2O 还原的石墨烯复合材料(Cu_2O/rGO)。直径约 200nm 的 Cu_2O 纳米球由直径约为 20nm 的纳米粒子组成,均匀分散在石墨烯表面和夹层上,在 Cu_2O/rGO 片中没有观察到团聚。通过甲基橙降解评估合成的 Cu_2O/rGO 复合材料。与纯 Cu_2O 纳米球相比,Cu_2O/rGO 复合材料显示出显著的甲基橙光催化降解效率。在紫外光照射 60min,96.6% 的初始甲基橙染料被 Cu_2O/rGO 复合物分解,然而,在相同条件下只有 52.6% 的初始染料被纯 Cu_2O 纳米球分解。光催化降解效率的提高可归因于 rGO 不仅作为电荷受体促进光生载流子的分离和转移,而且作为稳定 Cu_2O 和吸附水溶液中甲基橙分子的载体。Pu 等人[20]通过简便绿色的光化学反应制备了 Cu_2O/rGO 纳米异质结构。使用适量的 rGO,实现了电子从 Cu_2O 到 rGO 的快速转移,导致样品中的电荷分离越来越明显。发现 Cu_2O/rGO 纳米异质结构的光催化活性超过原始 Cu_2O 纳米簇、商业 Cu_2O 粉末和N掺杂的P-25 TiO_2 粉末。Zou 等人[21]合成了具有可调 Cu_2O 晶面（{111}、{110}和{100} 晶面）的 Cu_2O/rGO 复合材料，并用于在可见光下降解亚甲基蓝。图 6.5 为制备的 Cu_2O/rGO 复合材料的 SEM 表征结果，从图中可见，Cu_2O 颗粒均匀地分散在rGO 表面。结果表明，复合材料的催化行为与电子结构和界面连接有关。原位 ESR 研究表明，在 Cu_2O/rGO 复合材料上存在更多数量和更长寿命的超氧自由基 ($O_2^{-\cdot}$)。因此，亚甲基蓝在可见光下的降解性能排序为：八面体 Cu_2O{111}/rGO > 十二面体 Cu_2O{110}/rGO > 立方体 Cu_2O{100}/rGO。Zhang 等人[22]制备了立方 Cu_2O/rGO 并用于在可见光下去除甲基橙。与纯立方 Cu_2O 颗粒相比，所制备的立方 Cu_2O/rGO 纳米复合材料对甲基橙染料表现出增强的光催化活性，100min 内降解率为 100%，这主要归因于电荷传输增加，来自空位的光电子有效分离以及接触面积的改善。对于实际应用，将纳米结构的光催化剂固定在某些固体表面上进行分离和回收是理想的。Cai 等人[23]使用葡萄糖作为还原剂和交联剂，通过一锅水热法合成了 Cu_2O/rGO 复合气凝胶。结果表明，在 Cu_2O/rGO 复合气凝胶存在下，辐照 5h 后近 70% 的甲基橙被降解，而在 Cu_2O 纳米颗粒存在下，只有不到 30%的甲基橙被降解。这种改进可归因于以下因素：首先，rGO 是一种具有良好导电性的材料，

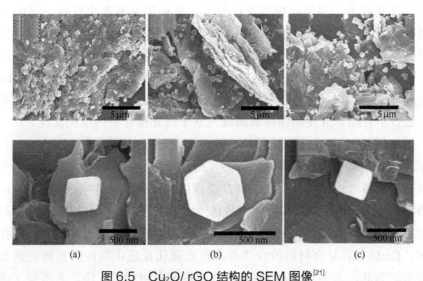

图 6.5 Cu_2O/rGO 结构的 SEM 图像[21]

（a）～（c）为不同视角下的图像

rGO 的掺入可以促进电荷转移并有助于分离 Cu_2O 上的光生电荷载流子；其次，Cu_2O/rGO 复合气凝胶在水性反应体系表面的漂浮可以保证其比 Cu_2O 粉末吸收更多的可见光；此外，所制备的 Cu_2O/rGO 对污染物表现出更高的吸附性，这是实现优异光催化的先决条件。

为此，本书对 Cu_2O 复合碳材料的光催化性能也进行了深入研究。

介孔碳是一种发展极其迅速的研究材料之一，其中氮掺杂的有序介孔碳（N-CMK-3）是指在介孔碳表面或者骨架中引入氮杂原子，使其表面性质发生变化，从而改善材料的浸润性和导电性[24,25]。研究者[26]认为氮元素上的孤对电子为碳的 sp^2 杂化结构提供额外的负电荷，起到载流子的作用，可以改善介孔碳材料的电子云分布和传输特性，使其导电性能增强。多孔碳表面的含氮官能团能改善材料的亲水性能，提高其与体系的相容性，拓宽应用范围。正六边形网格中的碳原子被 N 取代后会增加介孔碳材料的缺陷数量，孤对电子对金属阳离子的吸引力有利于金属颗粒在氮掺杂介孔碳内外表面的成核和锚定，有效地提高催化活性。通过对有序介孔碳进行改性，使其结构中含有氮杂原子或氮官能团，在参与多相催化反应时可以加强载体与活性组分间的相互作用、控制活性组分的粒子大小和分散度，进而提高催化剂的稳定性。同时，由于介孔碳表面存在孤对电子，能够加速催化系统中的电子传输及活性组分的活化。另外，氮原子的引入可生成新的活性位点，进而能够较好地与金属或者金属氧化物等进行复合，从而提高其催化性能。有序介孔碳材料负载铂纳米粒子制备出 Pt/CMK-3 催化剂可以提高催化剂的稳定性，增强其催化活性[27]；有序介孔碳负载金纳米微粒复合材料热稳定性明显提高[28]；在有序介孔碳框架中同时加入氮和硫可以作为可再生乙醇燃料电池的无金属催化剂[29]。基于以上优点，本书对 Cu_2O/N-CMK-3 复合光催化剂的制备进行了介绍，并使用四环素溶液考察了该光催化剂的催化性能。

另外，不同形貌的石墨烯材料已经被广泛研究，包括氧化石墨烯（GO）、还原氧化石墨烯（rGO）和膨胀石墨（EG）。尤其是半导体/石墨烯相关复合材料的制备吸引了大量研究者展开工作以提高光催化效率[30-33]。在各种石墨烯材料中，EG 通常用作催化剂载体[34]。EG 因其用作合成各种材料的载体而备受关注，包括 TiO_2/EG [35]、RuO_x/EG [36] 和 MnO_2/EG [37]。据我们所知，Cu_2O/EG 复合材料的合成及其在光催化反应中的应用直到现在还没有报道。为此，本书对 Cu_2O/EG 复合材料制备进行了介绍，并考察了该光催化剂的催化性能。

碳纳米管由于具有独特的一维结构、大的比表面积、超强的机械性能、高的化学和热稳定性以及良好的导电能力（功函数为4.18eV，是电子的良好受体），作为催化剂和催化剂载体受到了广泛的关注。根据管壁石墨层层数，碳纳米管可分为单壁碳纳米管和多壁碳纳米管。由于合成方法和结构的差异，碳纳米管可以分别呈现出导体或半导体的性质。碳纳米管具有独特结构和物化性质。半导体光催化剂与碳纳米管相结合形成异质结，具有多重优势。首先，光致电子能够沿碳纳米管一维方向快速传递，有利于减少光生电势对电荷分离的抑制作用从而提高量子效率。其次，某些CNTs可以看作一种窄带隙半导体材料，因此有希望扩展材料对太阳光的吸附范围，即作为光敏剂使用[38-44]。再次，利用碳纳米管具有大比表面积的特点，可以对污染物和降解中间产物进行富集，使降解物充分与光催化剂接触，从而提高降解效率并减少二次污染。例如$TiO_2/CNTs$[45]，$ZrO_2/CNTs$[46]，$CdS/CNTs$[47]等。在本书中，我们介绍了$Cu_2O/CNTs$复合材料的制备并进行了该催化剂的光催化性测试。

6.2 KAPs-B/Cu_2O 光催化剂光催化性能研究

超交联苯基聚合物（KAPs-B）是超交联聚合物的一种，首先通过一步沉淀法将其负载在不同量的Cu_2O上，并通过在可见光下降解甲基橙（MO）评价了催化剂的光催化性能[10]。

6.2.1 KAPs-B/Cu_2O 光催化剂的表征分析

表6.1和图6.6为催化剂的比表面积（BET）表征和对比。如图和表所示，纯KAPs-B和Cu_2O的比表面积分别为983.00m^2/g和8.09m^2/g。很明显，Cu_2O的比表面积非常低，而KAPs-B具有高比表面积微孔结构。通过将KAPs-B负载在Cu_2O上，得到的催化剂（质量分数为7% KAPs-B/Cu_2O）比纯Cu_2O具有更大的比表面积（49.70m^2/g），这有利于增强对有机污染物的吸附。

表 6.1　样品的比表面积

样品	比表面积/ (m²/g)
KAPs-B	983.00
7% KAPs-B/Cu$_2$O	49.70
Cu$_2$O	8.09

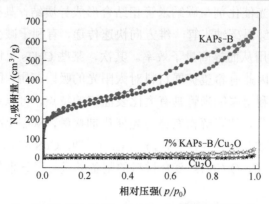

图 6.6　KAPs-B、Cu$_2$O 和 7% KAPs-B/Cu$_2$O（质量分数）的 N$_2$ 吸附-解吸等温线

图 6.7 为不同负载量的 KAPs-B/Cu$_2$O 催化剂的 XRD 光谱，其中，2θ 为 29.3°、36.2°、42.1°、61.3°和 73.4°的峰属于 Cu$_2$O{110}、{111}、{200}、{220} 和 {311} 晶面[48]，没有发现其他铜物种（Cu 和 CuO）相对应的其他峰，复合材料中仅存在纯 Cu$_2$O。此外，由于 KAPs-B 是无定形材料，因此在合成过程中，Cu$_2$O 的形成不受 KAPs-B 存在的影响。

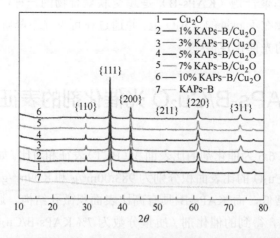

图 6.7　KAPs-B、Cu$_2$O 和 KAPs-B/Cu$_2$O 的 XRD 光谱

所制备的 KAPs-B/Cu₂O 催化剂的形态和 TEM 图像如图 6.8(a)~(e) 所示。从图中可以看出，所有样品中的 Cu_2O 均具有菱方十二面体结构，平均粒径约为 200~250nm。此外，随着制备过程中 KAPs-B 用量的增加，Cu_2O 表面的 KAPs-B 负载量增加。HRTEM 图像［图 6.8（f）］表明 KAPs-B 与 Cu_2O 表面结合得很好。

图 6.8　KAPs-B/Cu₂O 的 SEM[(a)~(e)]和 HRTEM 图像 (f)
（a）1% KAPs-B/Cu₂O；（b）3% KAPs-B/Cu₂O；（c）5% KAPs-B/Cu₂O；（d）7% KAPs-B/Cu₂O；
（e）10% KAPs-B/Cu₂O；（f）KAPs-B/Cu₂O

KAPs-B、Cu_2O 和 KAPs-B/Cu₂O 催化剂的化学结构由 FTIR 分析提供。从图 6.9 可以看到，对于 KAPs-B，在 1450~1600cm^{-1} 区域内芳环骨架拉伸的特征峰是明显的。对于 Cu_2O，在 623cm^{-1} 处观察到的吸收峰对应于 Cu(Ⅰ)—O 键的特征峰振动而在 3420cm^{-1} 处观察到的尖峰对应于羟基(—OH)的伸缩振动。这些结果表明 KAPs-B 成功地负载在 Cu_2O 颗粒上。

图 6.10 显示了不同 KAPs-B 负载量的 KAPs-B、Cu_2O 和 KAPs-B/Cu₂O 催化剂的 UV-Vis DRS 光谱。可以看到纯 Cu_2O 有很强的吸收，吸收边缘在 630 nm 左右，但 KAPs-B 对可见光没有明显的吸收峰。此外，随着 KAPs-B 负载量的增加，与纯 Cu_2O 相比，KAPs-B/Cu₂O 催化剂的吸收边缘几乎没有红移，但吸光度略有减弱。因此，KAPs-B/Cu₂O 催化剂保持了 Cu_2O 的主要光吸附性能，使其可以在可见光照射下工作。

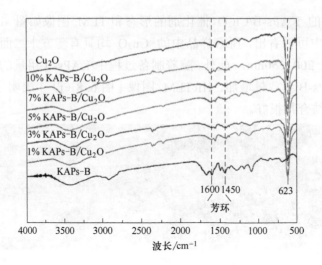

图 6.9　KAPs-B、Cu_2O 和 KAPs-B/Cu_2O 的 FTIR 光谱

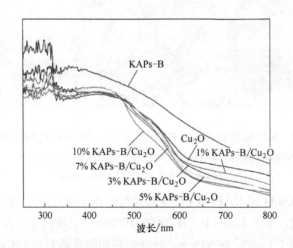

图 6.10　KAPs-B、Cu_2O 和 KAPs-B/Cu_2O 的 UV-Vis DRS 光谱

为了进一步研究电荷分离的效率，进行了光电流响应和电化学阻抗谱（EIS）的测量，结果如图 6.11 所示。从图 6.11（a）中可以看出，7% KAPs-B/Cu_2O 催化剂获得了更高的电流密度，这意味着其中更大的光生电子-空穴对分离率[49,50]。从图 6.11（b）中可以看出，7% KAPs-B/Cu_2O 催化剂的圆半径比纯 Cu_2O 的小，表明电荷载流子传输速度和有效电荷分离更快，进一步证实了 KAPs-B 的加载可以有效地促进光降解过程[51]。

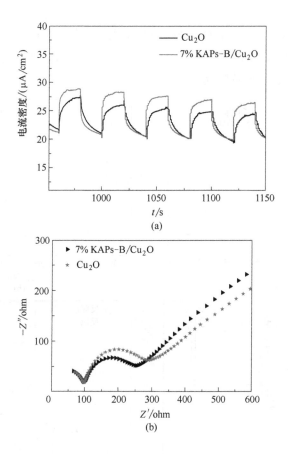

图 6.11 Cu$_2$O 和 KAPs-B/Cu$_2$O 复合物的光电流响应(a)和 EIS 光谱(b)

图 6.12 比较了 7%KAPs-B/Cu$_2$O 催化剂在 5 次循环测试前后的 XRD 谱、UV-vis DRS 谱、XPS 谱。可以看到，催化剂在 5 次循环测试前后的 XRD 谱图 [图 6.12（a）] 几乎保持不变，而 UV-vis DRS 谱图 [图 6.12（b）] 略有变化。图 6.12（c）中给出的 XPS 测量光谱表明存在 Cu、O 和 C，并且催化剂光谱在 5 次循环测试前后几乎保持不变。铜的化学状态如图所示。可以看到，反应前催化剂的 Cu 2p XPS 光谱在 932.2eV 和 951.9eV 处出现两个主峰，分别对应于 Cu$^+$ 的 Cu 2p$_{1/2}$ 和 Cu 2p$_{3/2}$[52]。从 Cu LMM 光谱可以看到，在 918.6eV 处没有观察到 CuO 物种的峰；在反应前后催化剂中没有形成铜单质。可以得出结论，反应前催化剂中没有 CuO 和 Cu 物种，Cu$_2$O 是唯一的物种。然而，MO 溶液光催化降解 5 次循环后，检测到 934.1eV 的小峰，这归因于 CuO 中的 Cu^{2+}，表明 KAPs-B/Cu$_2$O 光催化降解 5

次循环后表面只有少量 Cu_2O 转变为 CuO。然而,应该注意的是,生成的 CuO 量因为太少而无法在 XRD 光谱中观察到。以上结果表明,与 KAPs-B 结合后,Cu_2O 的稳定性明显提高。

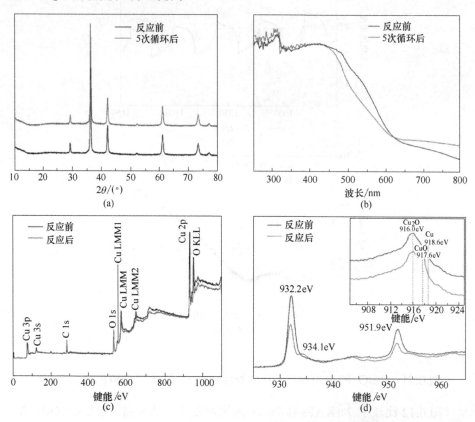

图 6.12 反应前后 7% KAPs-B/Cu_2O 催化剂的表征

(a)XRD 光谱; (b) UV-vis DRS 光谱; (c) 总谱; (d) Cu 2p XPS 光谱(插图为 Cu LMM XPS 光谱)

6.2.2 光催化性能测试

在可见光下,在不同比例的 KAPs-B/Cu_2O 复合材料上进行了甲基橙的光催化降解。如图 6.13 所示,在黑暗中进行 50min 的吸附-解吸平衡试验后,甲基橙浓度显著下降,表明溶液中大量的甲基橙被 Cu_2O 颗粒表面的微孔

KAPs-B 吸附。在可见光照射下，随着 Cu_2O 颗粒上的初始 KAPs-B 的质量分数从 1% 增加到 7%，光催化活性明显提高。然而，当 KAPs-B 的质量分数大于 7% 时，光催化降解效率没有显著提高。对于质量分数 7% KAPs-B/Cu_2O 催化剂，可以看到大约 92%的甲基橙在 60min 内分解。

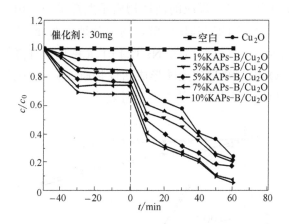

图 6.13　MO 的光催化降解曲线

光催化剂的稳定性是应用过程中的一个重要因素。图 6.14 显示了 7% KAPs-B/Cu_2O 对甲基橙降解循环 10 次的催化稳定性。如图 6.14 所示，在催化剂重复使用 10 次循环后，甲基橙的光催化转化率保持在大约 88.6%。

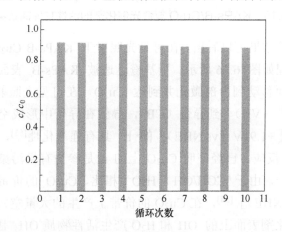

图 6.14　7%KAPs-B/Cu_2O 催化降解 MO 重复实验结果

6.2.3 光催化活性组分测试及机理研究

如图 6.15 所示，添加异丙醇（IPA）在甲基橙降解过程中没有带来明显的变化，表明·OH 作为参与反应的活性物质不发挥重要作用。然而，添加草酸铵（AO）时只有轻微的甲基橙降解，这清楚地表明 h^+ 是参与甲基橙降解的主要物种。此外，还可以看出在 N_2 吹扫下光催化活性降低，这是由于在光降解过程中，O_2 的短缺导致无法形成活性物质 $O_2^{-·}$。因此，$O_2^{-·}$ 和 h^+ 是 KAPs-B/Cu_2O 光催化剂体系中的主要活性物质。

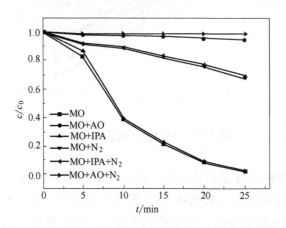

图 6.15 KAPs-B/Cu_2O 复合光催化剂的活性物种捕获实验

从这些实验结果可以看出，在可见光照射下 KAPs-B/Cu_2O 上甲基橙的光催化降解机理如图 6.16 所示。甲基橙迅速被 KAPs-B 表面吸收，然后通过微通道从表面移动到内部微孔并到达 Cu_2O。在可见光照射下，Cu_2O 的电子可以从价带（VB）到导带（CB），导致在导带中形成空穴。生成的具有更高正电位 [+1.91V（vs NHE）] 的 h^+ 具有强氧化性[53]，可以直接与吸附的 MO 发生反应。已经证明 Cu_2O 上的 h^+ 是参与有机污染物光降解的主要活性物质之一。由于与·OH/OH^-（H_2O）相比，Cu_2O 的价带电位不是正的 [+2.38V（vs NHE）][54]，在 Cu_2O 的价带上产生的大量空穴 h^+ 几乎不会氧化吸附在催化剂表面上的 OH^- 和 H_2O 产生活性物质·OH。因此，·OH 不应成为光催化过程的重要贡献者。此外，Cu_2O 异带上的电子可以被 KAPs-B 分子接收，苯环中的电子分布重新排列，导致电子结构的重叠，光催化剂的导电性增强。电子可以与吸附的 O_2 生成活性物种 $O_2^{-}·$，因为与 $O_2/O_2^{-}·$ [-0.046V（vs NHE）] 相比，Cu_2O 的导带电位（-0.26 V）更负[55]。活性物质 $O_2^{-}·$ 具有高氧化活性，可使甲基橙降解。基于以上分析，可以认为光生 h^+ 和 $O_2^{-}·$ 是参与甲基橙降解的主要活性物质,它也与活性物种捕获实验一致。由于生成的 $O_2^{-}·$ 和 h^+ 可以与吸附的甲基橙有效反应，因此会降低催化剂中光生电子和空穴的复合率。因此，该催化剂的光催化效率得以明显提高。

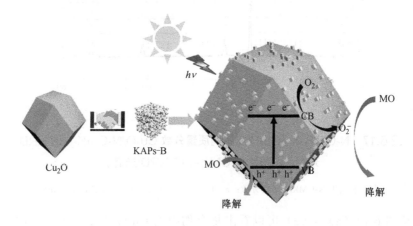

图 6.16 KAPs-B/Cu_2O 的光催化降解甲基橙机理

6.3 Cu_2O 复合碳材料的研究

6.3.1 N-CMK-3 光催化剂光催化性能研究

6.3.1.1 N-CMK-3 光催化剂的表征分析

如图 6.17 所示，观察到的峰值在 2θ 为 29.3°、36.2°、42.1°、61.3°和 73.4°对应于 Cu_2O 的 {110}、{111}、{200}、{220} 和 {311} 晶面，此外，峰值在 2θ 为 23.3°对应于 N-CMK-3 的 {002} 衍射峰也存在于复合物中[56]，说明在合成过程中 N-CMK-3 的添加并没有影响 Cu_2O 十二面体晶型的形成。以 Cu_2O 的 {111} 晶面为基准，用 Scherrer 方程 $[D=0.89\lambda/(\beta\cos\theta)]$ 来计算 Cu_2O 结晶的平均值样本的大小。通过计算可得 Cu_2O 的平均晶粒尺寸约 272 nm。

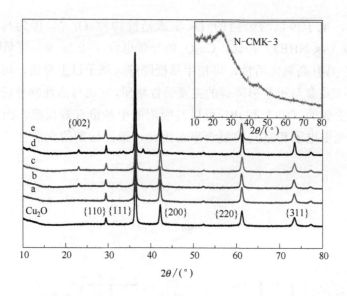

图 6.17　N-CMK-3、Cu_2O 和不同质量分数 Cu_2O/N-CMK-3 的 XRD 光谱（插图为 N-CMK-3 的 XRD 光谱）

复合材料中 N-CMK-3 质量分数：a—1%；b—3%；c—5%；d—7%；e—10%

从图 6.18（a）～（e）可以看出复合物中的 Cu_2O 为菱方十二面体结构，其粒径约为 200~300nm，这与 XRD 的估算结果基本一致。此外，我们还通过图 6.18（f）的 TEM 光谱观察到 N-CMK-3 均匀的包覆在 Cu_2O 颗粒的表面，说明 Cu_2O 与 N-CMK-3 在催化剂中结合良好。

从图 6.19 可以看出，Cu_2O 的光吸收边缘约在 630nm，掺杂 N-CMK-3 后可以提高 Cu_2O 催化剂的吸光度。这种新型催化剂可以获得更多的可见光，从而具有更高的光催化性能。

电化学性能被用来评价电子-空穴对在光催化剂中的分离和转移能力。因此，测量了 Cu_2O 和 Cu_2O/N-CMK-3 复合材料的瞬态光电流时间曲线，结果如图 6.20 所示。可以看出，掺杂不同 N-CMK-3 量的 Cu_2O/N-CMK-3 复合材料的光电流值明显高于纯 Cu_2O，进而可以提高其光催化性能。

图 6.21 为光催化剂的 XPS 谱图，图 6.21（a）为 Cu_2O/N-CMK-3 的总谱，图 6.21（b）为 Cu 2p 在 932.46eV 和 952.20eV 处的峰值，分别对应于 Cu $2p_{3/2}$ 和 Cu $2p_{1/2}$ 的光谱。图中特征峰值出现在 932.46eV 和 952.20eV，表明 Cu_2O 存在于光催化剂中。

图 6.18 Cu₂O/N-CMK-3 的 SEM[(a)~(e)]和 TEM 图像(f)

(a) 1% N-CMK-3/Cu₂O；(b) 3% N-CMK-3/Cu₂O；(c) 5% N-CMK-3/Cu₂O；(d) 7% N-CMK-3/Cu₂O；
(e) 10% N-CMK-3/Cu₂O；(f) 7% N-CMK-3/Cu₂O

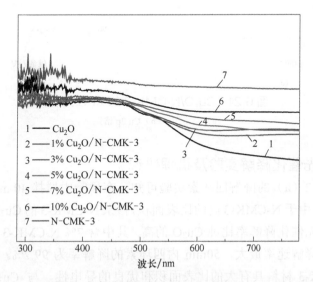

图 6.19 N-CMK-3、Cu₂O 和 Cu₂O/N-CMK-3 的 UV-Vis DRS 光谱

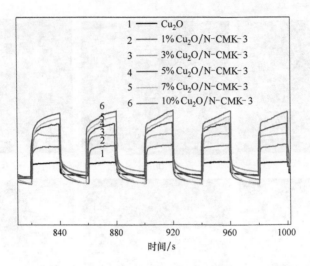

图 6.20 Cu₂O 和 Cu₂O/N-CMK-3 的光电流谱图

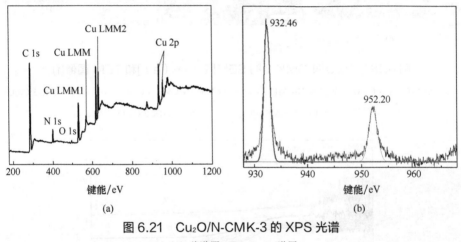

图 6.21 Cu₂O/N-CMK-3 的 XPS 光谱
（a）总谱图；（b）Cu 2p 谱图

6.3.1.2 光催化降解实验及机理研究

从图 6.22（a）的降解四环素实验可知，在黑暗中搅拌 40 min 后溶液达到吸附平衡。由于 N-CMK-3 高的比表面积，掺杂 N-CMK-3 的 Cu_2O/N-CMK-3 复合材料的光催化降解率比纯 Cu_2O 的高，其中含 7% N-CMK-3 的复合材料对四环素的降解速率最大，50min 内四环素的降解率为 99.2%。综上所述，由于 N-CMK-3 材料具有大的比表面积和优良的导电性，与 Cu_2O 结合后可以增加催化剂的比表面积，并且使电子可以快速转移到 N-CMK-3 材料的表面，进而增强电荷-空穴对的有效分离，增加光催化性能。图 6.22（b）为不同比例催化剂的降解速率图，对于该反应来说，符合准一级反应动力学，如图所示，该曲线具有较好的线性相关系数，其中 7% N-CMK-3 的反应速率常数 k =0.0851min^{-1}，为纯 Cu_2O 的反应速率常数 k =0.0456min^{-1} 的 1.87 倍。进而说明复合 N-CMK-3 材料后可以增强催化剂的电荷-空穴对的有效分离，从而增加光催化性能。

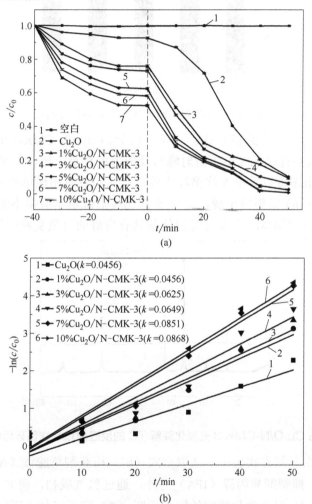

图 6.22　N-CMK-3、Cu_2O 和 Cu_2O/N-CMK-3 可见光光催化活性(a)和降解速率(b)

图 6.23 为 Cu_2O/N-CMK-3 复合光催化剂进行 5 次循环的 TC 降解图。从图 6.23 可以看出，在 5 次循环后，该光催化剂的活性仍然保持在 92.3%。

结果表明，与 N-CMK-3 相结合可以显著提高 Cu$_2$O 光催化剂的稳定性。

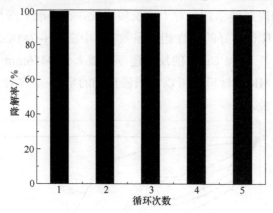

图 6.23　Cu$_2$O/N-CMK-3 光催化重复性实验

图 6.24 为反应前的和 5 次循环实验后 7% Cu$_2$O/N-CMK-3 的 XRD 图谱，5 次循环实验后样品的 Cu$_2$O 衍射峰与反应前的样品一致，但是衍射峰强度有所降低。由图可以看出 5 次循环实验后样品中没有出现单质 Cu 的衍射峰但有微弱的 CuO 衍射峰出现，这与 5 次循环后降解率有微小的下降结果一致。表明 Cu$_2$O/N-CMK-3 复合光催化剂具有良好的可重复利用性稳定性。

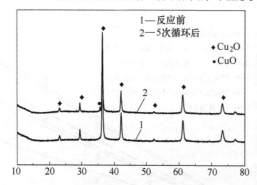

图 6.24　7% Cu$_2$O/N-CMK-3 光催化降解 TC 的反应前和 5 次循环后的 XRD 图谱

在活性组分测定实验中，选取空穴（h$^+$）捕获剂草酸铵（AO）、羟基自由基（·OH）抑制剂异丙醇（IPA）另外，通过氮气吸扫，研究 O$_2$ 生成超氧自由基（O$_2^-$·）对光催化降解的影响。图 6.25 展示了加入不同自由基捕获剂的 Cu$_2$O/N-CMK-3 催化剂对降解 TC 的影响。光反应时间为 50min 时，如图所示，加入 AO 和 IPA 对光催化降解影响显著，说明 h$^+$ 和 ·OH 为主要反应活性，我们通过通入氮气保护来验证电子是否能与被吸附的氧分子发生反应产生超氧自由基（$O_2^-·$），结果可知氮气氛围对降解结果影响也比较显著，说明电子能与氧分子形成超氧负离子自由基（$O_2^-·$）。

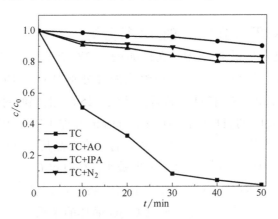

图 6.25　添加各种捕获剂对 Cu_2O/N-CMK-3 光催化降解四环素的影响

光催化机理如图 6.26 所示。Cu_2O 在可见光照射下，将电子从价带激发到导带，在价带中产生空穴。电子被激发到 Cu_2O 导带，然后电子可以迁移到 N-CMK-3 表面与被吸附的氧分子发生反应，产生超氧自由基（$O_2^-·$），空穴可与表面的 OH^- 或水发生反应，产生羟基自由基（·OH），二者都具有较高的氧化活性，可将四环素降解为 CO_2 和 H_2O。Cu_2O/N-CMK-3 不仅能够促进光生空穴和光生电子的分离，而且能有效地使 Cu_2O 内部载流子的复合重组，进而大大增强了复合物的光催化性能。

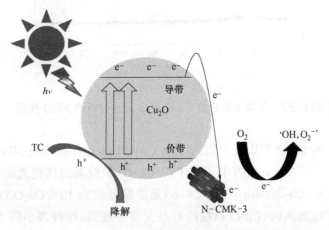

图 6.26　可见光下 Cu_2O/N-CMK-3 的光催化降解机理

6.3.2 Cu₂O/EG 光催化剂光催化性能研究

6.3.2.1 Cu₂O/EG 光催化剂的表征分析

Cu$_2$O/EG 复合材料的 XRD 光谱如图 6.27 所示。在 2θ 值 29.3°、36.2°、42.1°、61.3° 和 73.4° 处观察到的峰分别对应于 Cu$_2$O 的{110}、{111}、{200}、{220} 和 {311} 晶面。位于 25.6° 的强峰对应于 EG 的特征峰[57]。基于 Cu$_2$O 的{111}面，使用 Scherrer 方程 $[D=0.89\lambda/(\beta\cos\theta)]$ 计算样品的平均微晶尺寸。Cu$_2$O 具有最大的平均微晶尺寸（约 113nm）。当 Cu$_2$O 沉积在 EG 中时，其晶粒尺寸出现减小。Cu$_2$O/EG 的平均晶粒尺寸在 24~53nm。可能的原因是 EG 的存在可能会影响 Cu$_2$O 的结晶行为，随着 EG 的增加，Cu$_2$O 晶体将变得分散并抑制 Cu$_2$O 晶体的生长。

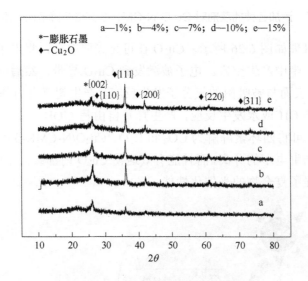

图 6.27 不同质量分数 Cu$_2$O/EG 复合材料的 XRD 光谱

图 6.28 为 Cu$_2$O 和质量分数为 10% Cu$_2$O/EG 复合材料的 SEM 图像。从图 6.28（a）可以看出，菱方十二面体的 Cu$_2$O 颗粒表面比较光滑，Cu$_2$O 的平均粒径约为 100~200nm。图 6.28（b）是质量分数为 10% Cu$_2$O/EG 的 SEM 图像，当将 EG 加入时，Cu$_2$O 颗粒的形态变得不规则，粒径减小到 20~50nm。这些结果表明 EG 的存在会影响 Cu$_2$O 的结晶行为，从而导致晶粒尺寸的减

小。从图 6.28（b）可以看出，大量的 Cu_2O 颗粒分布在 EG 片的表面上。这种结构为 EG 和 Cu_2O 颗粒之间提供了显著的比表面接触机会，并使其具有良好的载流子传输潜力。

图 6.28 Cu_2O (a) 和 10%Cu_2O/EG（质量分数）(b) 的 SEM 图像

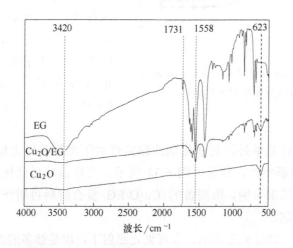

图 6.29 EG、Cu_2O/EG 和 Cu_2O 的 FTIR 谱图

EG、Cu_2O/EG 和 Cu_2O 的 FTIR 光谱如图 6.29 所示。EG 的光谱表明 $3420cm^{-1}$ 处的强峰可以归因于—OH 伸缩振动。在光催化反应中，反应活性与催化剂表面—OH 基团的数量密切相关，因为—OH 基团可以捕获光生空

穴（h^+）并转化为活性·OH 自由基。·OH 自由基是分解有机分子的主要自由基[58]。1731cm^{-1} 的谱带对应位于 EG 片材边缘的 COOH 基团的 C=O 拉伸，1558cm^{-1} 处的带可归因于 EG 片材的 C—C 的骨架振动[59]。在 Cu_2O 的 FT-IR 光谱的情况下，623cm^{-1} 处的吸收带可归因于 Cu（Ⅰ）—O 键的振动[60]。在 Cu_2O/EG 复合材料中观察到 EG 和纯 Cu_2O 的所有特征峰，表明 Cu_2O 已成功负载到 EG 表面。

6.3.2.2 光催化降解甲基橙性能测试及机理研究

Cu_2O/EG 复合材料的光催化活性主要通过可见光照射下甲基橙的光催化降解来评价。Cu_2O/EG 复合材料的光催化活性如图 6.30 所示。Cu_2O/EG 复合材料表现出比纯 Cu_2O 更好的光催化活性。随着 Cu_2O 质量分数从 1%变为 15%，甲基橙的光催化降解率显著提高。当 Cu_2O 含量进一步增加（质量分数大于 10%）时，光催化活性没有明显提高，60min 内约 96.7%的甲基橙被分解。考虑到催化剂的成本和效益，确定 Cu_2O/EG 复合材料中的最佳 Cu_2O 含量为 10 %。

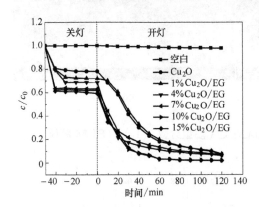

图 6.30 不同质量分数 Cu_2O/EG 复合材料光催化降解甲基橙的曲线

除了光催化效率外，光催化剂的稳定性对于实际应用也很重要。进行了 MO 循环降解实验，结果如图 6.31 所示。五次循环后催化剂的活性保持在约 92%。该结果表明，所制备的 Cu_2O/EG 复合材料可用作可见光下稳定高效的光催化剂。

光催化机理如图 6.32 所示。在可见光照射下，用足够多的光子照射 Cu_2O 会将电子从价带激发到导带，从而在价带中产生空穴。空穴可以与表面的氢氧根离子或水反应生成羟基自由基（·OH），它具有很高的氧化活性，能够降解甲基橙。同时，电子可以与吸附的分子氧反应产生超氧负离子自由基（$O_2^{-·}$），在水分子的存在下，它可以进一步形成高反应性的·OH。由于 EG 独特的二维共轭结构，它可以分离光生电荷对，进而降低空穴-电子对的复合率并增强其光活性[14]。此外，由于 EG 的比表面积较大，对甲基橙的吸附能力显著增大，从而增强其光催化活性。

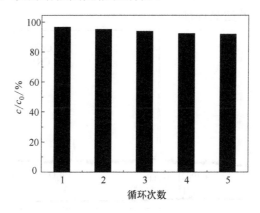

图 6.31 Cu$_2$O/EG 光催化重复性实验

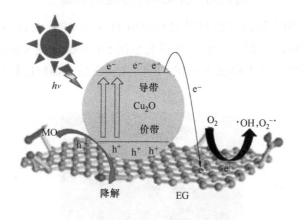

图 6.32 Cu$_2$O/EG 光催化降解甲基橙机理

6.3.3 Cu$_2$O/CNTs 复合材料光催化性能研究

6.3.3.1 Cu$_2$O/CNTs 复合材料的表征分析

图 6.33 为碳纳米管负载氧化亚铜的 XRD 表征结果，从图中可以看出，碳纳米管负载 Cu$_2$O 有几个明显的吸收峰，在 $2\theta=26.4°$ 有一个特征峰，且与碳纳米管的标准峰基本吻合，所以此峰为碳纳米管的吸收峰。在 2θ 约 30°~45°

中，有几处明显的吸收峰，29.3°、36.2°、42.1°、61.3°、73.4°分别对应晶体的{110}、{111}、{200}、{220}、{311}晶面。结果表明Cu_2O在CNTs载体表面上具有良好的结晶性能。

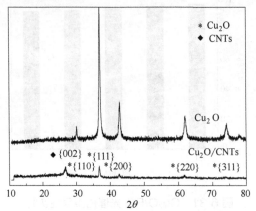

图6.33　Cu_2O/CNTs的XRD图

图6.34为用直接沉淀法制得的Cu_2O/CNTs的SEM图像，由图可以清晰地观察到，该样品主要是碳纳米管和大小在50~100nm的Cu_2O晶体组成，由于碳纳米管提供了大的比表面积，所以Cu_2O晶体颗粒比较均匀地分布于CNTs表面。

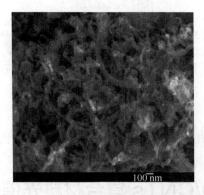

图6.34　Cu_2O/CNTs的SEM图像

6.3.3.2　Cu_2O/CNTs复合材料的性能测试和机理研究

图6.35是Cu_2O/CNTs催化脱色降解30mg/L甲基橙溶液的曲线图。从图中的五条曲线可以看出，含量为5%Cu_2O的碳纳米管对甲基橙溶液的催化效果相比较其他含量而言效果较佳。随着样品中Cu_2O含量的逐渐增大，纳

米氧化亚铜在碳纳米管上的负载量也随之逐渐增大。但实验结果表明，随着氧化亚铜的负载量持续增大，碳纳米管负载纳米 Cu_2O 样品对甲基橙溶液的催化效果升高的幅度不是很明显。因此，从负载量的角度考虑，负载氧化亚铜时 Cu_2O 的含量适宜控制在 5%左右。

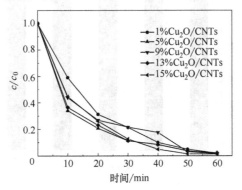

图 6.35　不同质量分数 Cu_2O/CNTs 复合材料对甲基橙降解率的影响

由图 6.36 可知，随着甲基橙浓度的增加，降解率逐渐降低。当甲基橙浓度为 10 mg/L 时，在 40 min 内降解率可达到 96.6%，说明初始浓度对光催化有较大的影响。随着甲基橙浓度的增大，溶液的颜色加深，进而降低了对光的吸收，使降解率降低。但是，随着光反应时间的增加，浓度较大的甲基橙溶液的降解率仍然会增加。

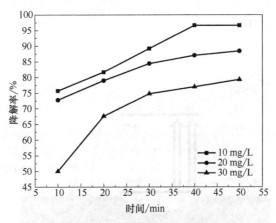

图 6.36　甲基橙浓度对降解率的影响

图 6.37 比较了 Cu_2O/CNTs 复合物的循环光催化活性。从图中可看出，经过 5 次循环后，Cu_2O/CNTs 的光催化活性逐渐降低，但其降解率仍能达到

90%以上。说明复合 CNTs 不仅提高了 Cu_2O 的光催化活性,也改善了催化剂的稳定性。

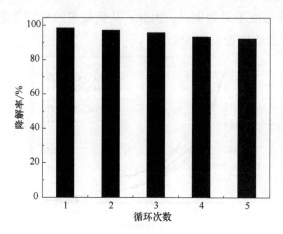

图 6.37　不同循环次数的光催化降解 MO 曲线

光催化机理如图 6.38 所示。Cu_2O 在可见光照射下,将电子从价带激发到导带,在价带中产生空穴。电子被激发到 Cu_2O 导带,然后电子可以迁移到 CNTs 表面与被吸附的氧分子发生反应,产生超氧负离子自由基($O_2^{-}·$),空穴可与表面的 OH^- 或水发生反应,产生羟基自由基(·OH),二者都具有较高的氧化活性,可降解甲基橙。$O_2^{-}·$ 和 ·OH 在可见光照射下均能有效地将甲基橙分解为 CO_2、H_2O 等。

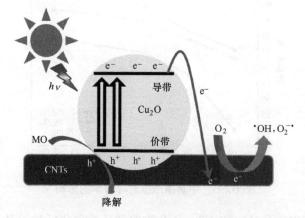

图 6.38　可见光下 Cu_2O/CNTs 光催化降解甲基橙机理

参考文献

[1] Tsyurupa M P, Blinnikova Z K, Borisov Y A, et al. Physicochemical and adsorption properties of hypercross-linked polystyrene with ultimate cross-linking density [J]. J Sep Sci, 2014, 37: 803-810.

[2] Tan L, Tan B. Hypercrosslinked porous polymer materials: design, synthesis and applications [J]. Chem Soc Rev, 2017, 46: 3322-3356.

[3] Jia Z, Wang K, Li T, et al. Functionalized hypercrosslinked polymers with knitted N-heterocyclic carbene-copper complexes as efficient and recyclable catalysts for organic transformations [J]. Catal Sci Technol, 2016, 6: 4345-4355.

[4] Wang K, Jia Z, Yang X, et al. Acid and base coexisted heterogeneous catalysts supported on hypercrosslinked polymers for one-pot cascade reactions [J]. J Catal, 2017, 348: 168-176.

[5] Jia Z, Wang K, Tan B, et al. Ruthenium complexes immobilized on functionalized knitted hypercrosslinked polymers as efficient and recyclable catalysts for organic transformations [J]. Adv Synth Catal, 2017, 359: 78-88.

[6] Liu S, Hu Q, Zheng J, et al. Knitting aromatic polymers for efficient solid-phase microextraction of trace organic pollutants [J]. J Chromatogr A, 2016, 1450: 9-16.

[7] Li B, Gong R, Wang W, et al. A new strategy to microporous polymers: knitting rigid aromatic building blocks by external cross-linker [J]. Macromolecules, 2011, 44: 2410-2414.

[8] Zhou K, Shi Y, Jiang S, et al. Facile preparation of Cu_2O/carbon sphere heterostructure with high photocatalytic activity [J]. Mater Lett, 2013, 98: 213-216.

[9] Ai Z, Xiao H, Mei T, et al. Electro-Fenton degradation of rhodamine B based on a composite cathode of Cu_2O nanocubes and carbon nanotubes [J]. J Phys Chem C, 2008, 112: 1192-11935.

[10] Luo Y S, Ren Q F, Li J L, et al. Synthesis and optical properties of multiwalled carbon nanotubes beaded with Cu_2O nanospheres [J]. Nanotechnology, 2006, 17: 5836.

[11] Yang L, Chu D, Wang L, et al. Synthesis and photocatalytic activity of chrysanthemum-like Cu_2O/Carbon Nanotubes nanocomposites [J]. Ceram Int, 2016,

42:2502-2509.

[12] Zhao Q, Wang J, Li Z, et al. Preparation of Cu_2O/exfoliated graphite composites with high visible light photocatalytic performance and stability [J]. Ceram Int, 2016, 42:13273-13277.

[13] Khare P, Singh A, Verma S, et al. Sunlight-induced selective photocatalytic degradation of methylene blue in bacterial culture by pollutant soot derived nontoxic graphene nanosheets [J]. ACS Sustain Chem Eng, 2018, 6: 579-589.

[14] Du W, Wu M, Zhang M, et al. High-quality grapheme films and nitrogen-doped organogels prepared from the organic dispersions of grapheme oxide [J]. Carbon, 2018,129: 15-20.

[15] Yan H, Wu H, Li K, et al. Influence of the surface structure of graphene oxide on the adsorption of aromatic organic compounds from water [J]. ACS Appl Mater Inter, 2015, 7: 6690-6697.

[16] Tu K, Wang Q, Lu A, et al.Portable visible-light photocatalysts constructed from Cu_2O nanoparticles and graphene oxide in cellulose matrix [J]. The J Phys Chem C, 2014, 118: 7202-7210.

[17] Nie J, Li C, Jin Z, et al. Fabrication of MCC/Cu_2O/GO composite foam with high photocatalytic degradation ability toward methylene blue[J]. Carbohyd Polym, 2019, 223:115101.

[18] Wang H L, Zhang L S, Chen Z G, et al. Semiconductor heterojunction photocatalysts: design, construction, and photocatalytic performances [J]. Chem Soc Rev 2014, 43, 5234-5244.

[19] Abulizi A, Yang G, Zhu J J. One-step simple sonochemical fabrication and photocatalytic properties of Cu_2O-rGO composites [J]. Ultrason Sonochem, 2014, 21: 129-135.

[20] Pu Y C, Chou H Y, Kuo W S, et al. Interfacial charge carrier dynamics of cuprous oxide-reduced graphene oxide (Cu_2O-rGO) nanoheterostructures and their related visible-light-driven photocatalysis [J]. Appl Catal B, 2017, 204: 21-32.

[21] Zou W, Zhang L, Liu L, et al. Engineering the Cu_2O-reduced graphene oxide interface to enhance photocatalytic degradation of organic pollutants under visible light[J]. Appl Catal B, 2016, 181: 495-503.

[22] Zhang W, Li X, Yang Z, et al. In situ preparation of cubic Cu_2O-RGO nanocomposi-

tes for enhanced visible-light degradation of methyl orange [J]. Nanotechnology, 2016, 27: 265703.

[23] Cai J, Liu W, Li Z, One-pot self-assembly of Cu$_2$O/RGO composite aerogel for aqueous photocatalysis [J]. A Surf Sci, 2015, 358: 146-151.

[24] Wang J, Ma R, Zhou Y, et al. A facile nanocasting strategy to nitrogen-doped porous carbon monolith by treatment with ammonia for efficient oxygen reduction [J]. J Mater Chem A, 2015, 3: 12836-12844.

[25] Xiao C, Chen X, Fan Z, et al. Surface-nitrogen-rich ordered mesoporous carbon as an efficient metal-free electrocatalyst for oxygen reduction reaction[J]. Nanotech, 2016, 27: 402-445.

[26] 刘宗. 氮掺杂碳材料的合成及其在催化反应中的应用 [D]. 青岛：青岛科技大学, 2019.

[27] 张巧利, 徐强, 张媛媛, 等. 磁性介孔碳的制备及对水体中染料的吸附去除 [J]. 环境化学, 2018, 37: 2548-2554.

[28] Benzigar M R, Talapaneni S N, Joseph S, et al. Recent advances in functionalized micro and mesoporous carbon materials: synthesis and applications [J]. Chem Soc Rev, 2018, 47: 2680-2721.

[29] Qiu Y, Huo J, Jia F, et al. N- and S- doped mesoporous carbon as metal-free cathode catalysts for direct biorenewable alcohol fuel cells [J]. J Mater Chem, 2016, 4: 83-95.

[30] Li H, Liu R, Liu Y, et al. Carbon quantum dots/Cu$_2$O composites with protruding nanostructures and their highly efficient (near) infrared photocatalytic behavior [J]. J Mater Chem, 2012, 22: 17470-17475.

[31] Tran P D, Batabyal S K, Pramana S S, et al. A cuprous oxide–reduced graphene oxide (Cu$_2$O-rGO) composite photocatalyst for hydrogen generation: employing rGO as an electron acceptor to enhance the photocatalytic activity and stability of Cu$_2$O [J]. Nanoscale, 2012, 4: 3875-3878.

[32] Zhao Q, Meng S M, Wang J L, et al. Preparation of solid superacid $S_2O_8^{2-}$/TiO$_2$-exfoliated graphite (EG) and its catalytic performance[J]. Ceram Int, 2014, 40: 16183-16187.

[33] An X Q, Li K F, Tang J W. Cu$_2$O/Reduced graphene oxide composites for the photocatalytic conversion of CO$_2$ [J]. Chem Sus Chem, 2014, 7: 1086-1093.

[34] Inagaki M, Suwa T. Pore structure analysis of exfoliated graphite using image

[35] Ndlovu T, Kuvarega A T, Arotiba O A, et al. Exfoliated graphite/titanium dioxide nanocomposites for photodegradation of eosin yellow [J]. Appl Surf Sci 2014, 300: 159-164.

[36] Mitra S, Lokesh K S, Sampath S. Exfoliated graphite-ruthenium oxide composite electrodes for electrochemical supercapacitors [J]. J Power Sources, 2008, 185:1544-1549.

[37] Yang YJ, Liu E H, Li L M, et al. Nanostructured MnO_2/exfoliated graphite composite electrode as supercapacitors [J]. J Alloys Compd, 2009, 487: 564-567.

[38] Reyhani A, Mortazavi S Z, Moshfegh A Z, et al. Enhanced electrochemical hydrogen storage by catalytic Fe-doped multi-walled carbon nanotubes synthesized by thermal chemical vapor deposition [J]. J Power Sources, 2009, 188: 404-410.

[39] Reyhani A, Mortazavi S Z, Mirershadi S, et al. H_2 adsorption mechanism in Mg modified multi-walled carbon nanotubes for hydrogen storage [J]. Int J Hydrogen Energy, 2012, 37: 1919-1926.

[40] Reyhani A, Mortazavi S Z, Mirershadi S, et al. Nozad Golikand, Hydrogen storage in decorated multiwalled carbon nanotubes by Ca, Co, Fe, Ni, and Pd nanoparticles under ambient conditions [J]. J Phys Chem C, 2011,115: 6994-7001.

[41] Afsharmanesh E, Karimi-Maleh H, Pahlavan A, et al. Electrochemical behavior of morphine at ZnO/CNT nanocomposite roomtemperature ionic liquidmodified carbon paste electrode and its determination in real samples [J]. J Mol Liq, 2013, 181:8-13.

[42] Karimi-Maleh H, Biparva P, Hatami M. A novel modified carbon paste electrodebased on NiO/CNTs nanocomposite and (9,10-dihydro-9, 10-ethanoanthracene-11,12-dicarboximido)-4-ethylbenzene-1,2-diol as a mediator for simultaneous determination of cysteamine, nicotin amide adenine dinucleotide and folic acid [J]. Biosens Bioelectron, 2013, 48: 270-275.

[43] Shahmiri M R, Bahari A, Karimi-Maleh H, et al. Ethynylferrocene–NiO/MWCNT nanocomposite modified carbon paste electrode as a novel voltammetric sensor for simultaneous determination of glutathione and acetaminophen[J]. Sens Actuators B, 2013, 177: 70-77.

[44] Ensafi A A, Bahrami H, Rezaei B, et al. Application of ionic liquid-TiO_2 nanoparticlemodified carbon paste electrode for the voltammetric determination of

benserazide in biological samples [J]. Mater Sci Eng C, 2013, 3: 831-835.

[45] Zhu L W, Zhou L K, Li H X, et al. One-pot growth of free-standing CNTs/TiO_2 nanofiber membrane for enhanced photocatalysis [J]. Mater Lett, 2013, 95: 13-16.

[46] Suárez G, Jang B K, Aglietti Esteban F. Fabrication of dense ZrO_2/CNT composites: influence of bead-milling treatment [J]. Metallurgical and Materials Transactions A, 2013, 44:4374-4381.

[47] Wang X X, Liu M H, Chen Q Y, et al. Synthesis of CdS/CNTs photocatalysts and study of hydrogen production by photocatalytic water splitting [J]. Int J Hydrogen Energy, 2013, 38: 13091-13096.

[48] Zou W, Zhang L, Liu L, et al. Engineering the Cu_2O–reduced graphene oxide interface to enhance photocatalytic degradation of organic pollutants under visible light [J]. Appl Catal B-Environ, 2016, 181: 495-503.

[49] Toe C Y, Zheng Z, Wu H, et al. Photocorrosion of cuprous oxide in hydrogen production: rationalising self-oxidation or self-reduction [J]. Angew Chem Int Edit, 2018, 130:13801-13805.

[50] Liu H, Tian K, Ning J, et al. One-step solvothermal formation of Pt nanoparticles decorated Pt^{2+}-doped α-Fe_2O_3 nanoplates with enhanced photocatalytic O_2 evolution [J]. ACS Catal, 2019, 9: 1211-1219.

[51] Etogo A, Liu R, Ren J, et al. Facile one-pot solvothermal preparation of Mo-doped Bi_2WO_6 biscuit-like microstructures for visible-light-driven photocatalytic water oxidation [J]. J Mater Chem A, 2016, 4: 13242-13250.

[52] He B, Liu R, Ren J, et al. One-step solvothermal synthesis of petalous carbon-coated Cu^+-doped CdS nanocomposites with enhanced photocatalytic hydrogen production [J]. Langmuir, 2017, 33: 6719-6726.

[53] Meng S, Li D, Sun M, et al. Sonochemical synthesis, characterization and photocatalytic properties of a novel cube-shaped $CaSn(OH)_6$ [J]. Catal Commun, 2011, 12:972-975.

[54] Ye L, Liu J, Jiang Z, et al. Facets coupling of Biobr-G-C_3N_4 composite photocatalyst for enhanced visible-light-driven photocatalytic activity [J]. Appl Catal B-Environ, 2013, 142:1-7.

[55] Bandyopadhyay A, Pal A J. Large conductance switching and memory effects in organic molecules for data-storage applications [J]. Appl Phys Lett, 2003, 82:

1215-1217.

[56] Wei W, Liang H, Parvez K, et al. Nitrogen-doped carbon nanosheets with size-defined mesopores as highly efficient metal-free catalyst for the oxygen reduction reaction [J]. Angew Chem In Ed, 2014, 53: 1570-1574.

[57] Hummers Jr W S, Offeman R E. Preparation of graphitic oxide [J]. J Am Chem Soc, 1958, 80: 1339.

[58] Li J, Wei L F, Yu C L, et al. Preparation and characterization of graphene oxide/Ag_2CO_3 photocatalyst and its visible light photocatalytic activity [J]. Applied Surface Science, 2015, 358: 168-174.

[59] Titelman G I, Gelman V, Bron S, et al. Characteristics and microstructure of aqueous colloidal dispersions of graphite oxide [J]. Carbon, 2005, 43: 641-649.

[60] Tian Y L, Chang B B, Fu J, et al. Graphitic carbon nitride/Cu_2O heterojunctions: preparation, characterization, and enhanced photocatalytic activity under visible light [J]. J Solid State Chem, 2014, 212:1-6.